Reviewer praise for *The Story of Life*

"Carroll's book has the potential to open [teachers'] eyes to the merits of story-based instruction when they see that their students are more engaged and have a stronger grasp of biological concepts."

—Rebecca Brewer, Troy High School, Troy, MI

"The short chapters are easy to understand, while still explaining meaty concepts, lending support to the idea that narratives are superior to expository texts when it comes to comprehension."

—Mary Ann Price, Columbia University

"I was very impressed with the overall strength of the stories and the diversity of the topics that they covered. I think this book is a perfect way to connect the science to the students and increase their interest in how science happens."

—Anna Bess Sorin, University of Memphis

"As [teachers], one of our main focuses for the last several years has been to find ways to supplement an area where students are struggling: reading. By having a non-textbook that covers biology content, we would be able to help achieve our goal. The chapters are [concise], engaging, relevant, and still manage to connect directly to the content."

—AJ Prawitz, John F. Hodge High School, St. James, MO

"Science comes out of creative and persevering personalities; this drama is not elaborated enough when disseminating science to the public. The narrative approach in this book is a sensible counternarrative that adds to the captivating story that is scientific discovery. Fascinating people generating fascinating facts."

—Dale Casamatta, University of North Florida

"*The Story of Life* is perfectly positioned to function as engaging and fascinating supplementary reading for students in a biology course."

—Paul Strode, Fairview High School, Boulder, CO

Reviewer praise for *The Story of Life*:

THE STORY

of

LIFE

Great Discoveries in Biology

THE STORY

of

LIFE

Great Discoveries in Biology

SEAN B. CARROLL

W. W. Norton & Company

New York • London

W. W. Norton & Company has been independent since its founding in 1923, when William Warder Norton and Mary D. Herter Norton first published lectures delivered at the People's Institute, the adult education division of New York City's Cooper Union. The firm soon expanded its program beyond the Institute, publishing books by celebrated academics from America and abroad. By midcentury, the two major pillars of Norton's publishing program—trade books and college texts—were firmly established. In the 1950s, the Norton family transferred control of the company to its employees, and today—with a staff of four hundred and a comparable number of trade, college, and professional titles published each year—W. W. Norton & Company stands as the largest and oldest publishing house owned wholly by its employees.

Editor: Betsy Twitchell
Associate Editor: Katie Callahan
Editorial Assistants: Taylere Peterson and
    Danny Vargo
Managing Editor, College: Marian Johnson
Managing Editor, College Digital Media: Kim Yi
Senior Production Manager: Sean Mintus
Media Editor: Kate Brayton
Associate Media Editor: Gina Forsythe
Media Project Editor: Jesse Newkirk

Media Editorial Assistant: Katie Daloia
Content Development Specialist: Todd Pearson
Marketing Manager, Biology: Stacy Loyal
Design Director: Lissi Sigillo
Book Designer: Jordan Wannemacher
Director of College Permissions: Megan Schindel
Senior Photo Editor: Ted Szczepanski
College Permissions Assistant: Patricia Wong
Composition: Westchester Publishing Services
Manufacturing: Transcontinental

Permission to use copyrighted material is included alongside the appropriate content including the following:

Page 126: Excerpt from Letter from Carl R. Woese to Francis Crick (June 24, 1969). Reprinted by permission of Robert Woese in his capacity as attorney-in-fact for Mrs. Gabriella H. Woese.

Chapter 13: Adapted from *The Serengeti Rules: The Quest to Discover How Life Works and Why It Matters* by Sean B. Carroll. Copyright © 2016 by Sean B. Carroll. Published by Princeton University Press. Reprinted by permission.

Chapter 8 is adapted from *Remarkable Creatures: Epic Adventures in the Search for the Origin of Species* by Sean B. Carroll (2009, Houghton Mifflin Harcourt).

Library of Congress Cataloging-in-Publication Data

Names: Carroll, Sean B., author.
Title: The story of life : great discoveries in biology / Sean B. Carroll.
Description: New York : W.W. Norton & Company, [2019] | Includes
    bibliographical references and index.
Identifiers: LCCN 2018046625 | ISBN 9780393631562 (paperback)
Subjects: LCSH: Biology—History.
Classification: LCC QH305 .C29 2019 | DDC 570—dc23 LC record available at
    https://lccn.loc.gov/2018046625

W. W. Norton & Company, Inc., 500 Fifth Avenue, New York, NY 10110
wwnorton.com
W. W. Norton & Company Ltd., 15 Carlisle Street, London W1D 3BS

1 2 3 4 5 6 7 8 9 0

In memory of Jack Repcheck and Jonathan Barker—
two great storytellers who taught me a lot,
and two of the finest people I have had the pleasure to know.

ABOUT THE AUTHOR

SEAN B. CARROLL is an evolutionary biologist, author, educator, and film producer. He is vice president for Science Education at the Howard Hughes Medical Institute, the Simon-Belo Professor of Biology at the University of Maryland, and professor emeritus at the University of Wisconsin–Madison. He has previously written several books including *The Serengeti Rules*, *Brave Genius*, *Endless Forms Most Beautiful*, *Into the Jungle*, and *Remarkable Creatures*, and is the executive producer of more than a dozen feature documentary films. He is a member of the National Academy of Sciences, the American Philosophical Society, and the American Academy of Arts and Sciences, and has received the Distinguished Service Award from the National Association of Biology Teachers.

Brief Contents

Contents

Storytelling is a universal and powerful way humans connect, but it is underutilized in science classrooms. Stories help us understand how science is done; the stories in this book were chosen to highlight some of the most important and astonishing discoveries in biology.

Two stories show how the scientific process works. The first demonstrates how a radical idea can eventually gain acceptance; the second underscores the serious consequences when the process is undermined.

Robin Warren* and Barry Marshall's* revolutionary discovery and unorthodox demonstration that *H. pylori* is the cause of ulcers.

Do vaccines cause autism? The story of the physician who raised the alarm about a link between vaccination and autism, and how that claim was tested and debunked, but affected vaccination rates across the world.

*Won Nobel Prize.

PART III EVOLUTION AND THE ORIGINS OF BIOLOGICAL DIVERSITY 77

Evolution is said to be the biggest idea humans ever conceived. The stories of several pioneers who illuminated key chapters in the story of life, and the origins of species, cells, animals, and humans.

PART IV ECOLOGY

How does nature work? Stories recounting experimental approaches to ecology that revealed surprising truths about the factors that shape ecosystems and the environment, and that are the scientific foundation for modern environmental stewardship.

PART V PHYSIOLOGY AND MEDICINE

How do the body's complex organs and systems function? The mysteries of development, immunity, and the brain have been penetrated by a few seminal experiments and discoveries, with profound implications for medicine.

17 Unleashing Potential

The surprising demonstration by John Gurdon* that mature somatic cells can be reprogrammed to generate a cloned animal, and Shinya Yamanaka's* discovery of the molecular recipe for reprogramming.

18 The Arsenal of Immunity

How does the body mount a specific immune response to foreign invaders? The story of Susumu Tonegawa's* discovery of the molecular basis of antibody diversity.

19 Everyone Has a Split Personality

How does the brain work? The story of Roger Sperry's* pioneering studies of human split-brain patients and his discovery of the different roles of left and right hemispheres.

Preface

WHY STORIES?

Tell me a fact and I'll learn. Tell me the truth and I'll believe.
But tell me a story and it will live in my heart forever.
—NATIVE AMERICAN PROVERB

Louis got an attack of the 'flu and retired to bed, and so it came about that on the morning of 17 July I went out by myself, with the two Dalmatians Sally and Victoria, to see what I could find of interest at nearby Bed I exposures. I turned my steps towards a site not far west of the junction of two gorges. . . . There was indeed plenty of material lying on the eroded surface . . . some no doubt as a result of the rains earlier that year. But one scrap of bone that caught and held my eye was not lying loose on the surface but projecting from beneath. It seemed to be part of a skull. . . . I carefully brushed away a little bit of the deposit, and then I could see parts of two large teeth in place in the upper jaw. They <u>were</u> hominid. It was a hominid skull, apparently in situ, and there was a lot of it there.

I rushed back to camp shouting out "I've got him! I've got him! I've got him!"

"Got what? Are you hurt?" Louis asked.

"Him, the man! Our man," I said. "The one we've been looking for. Come quick."

ONE WOMAN COMBING A HILLSIDE in Africa for scraps of bone does not arouse the stomach-churning fear of a soldier charging up a heavily fortified beach in France, or the nail-biting suspense of a piloted spacecraft approaching the surface of the moon. But this personal moment of discovery was just as great a turning point as the Normandy or moon landings, for it would also rewrite human history.

The story is Mary Leakey's account of the morning of Friday, July 17, 1959, when, at age 46, she found what she and her husband Louis had been seeking for 24 years—indisputable evidence that Africa was the cradle of humanity. Up until that moment, the question of the origin of humans was wide open, with most evidence and a great deal of cultural bias pointing away from Africa and toward Asia. The fragmented skull was the reward for the indefatigable efforts of these two brave pioneers who were willing to live most of their lives in the bush in the hope of finding something that would upend not only the scientific world, but society at large.

The story of Mary and Louis Leakey is one of the nineteen stories that make up *The Story of Life*. You may wonder: Why *stories* of life? And, why these stories in particular?

Those are good questions that I will address in a moment. But the essence of the rationale for this book is captured by a statement made by another pioneer whose discovery is chronicled here. With his partner Francis Crick, James D. Watson discovered the base pairing rules that were the key to the double helix structure, the function of DNA, and the mechanism of heredity. He gave an interview on the fiftieth anniversary of his and Crick's momentous discovery:

INTERVIEWER: What are you most proud of?

WATSON: My textbook *The Molecular Biology of the Gene* and my book *The Double Helix*.

INTERVIEWER: Not the actual discovery of the double helix?

WATSON: No, because the double helix was going to be found in the next year or two. It was just waiting to be found, and I was the one who finally found it because I was the most obsessed about it.

INTERVIEWER: Why did you choose to write a book focusing more on the people involved rather than the science of the double helix?

WATSON: I wanted to see if I could write a good book. It was ahead of its time, you could say, in terms of style. I wasn't thinking of myself as a scientist, you know. My heroes were never scientists. They were Graham Greene and Christopher Isherwood, you know, good writers.

Watson and Crick's discovery had earned the pair nearly instant worldwide fame and the 1962 Nobel Prize in Physiology or Medicine. But it wasn't that historic accomplishment that made Watson most proud, it was his telling of the story in his book *The Double Helix*!

Why would Watson feel this way?

He explained his reasons in the preface to what turned out to be a best-selling book. "As I hope this book will show, science seldom proceeds in the straightforward logical manner imagined by outsiders. Instead, its steps forward (and sometime backward) are often very human events in which personalities and cultural traditions play major roles." Watson wanted to convey "the spirit of adventure," but most importantly, he felt the story should be told because "there remains general ignorance about how science is 'done.'"

My main goals in writing this book are the same as Watson's—to offer a good book, made up of good stories, to convey the spirit of adventure, and to show how science is done. And while those may be sufficient incentives for some of you to open it, my rationale for writing *The Story of Life* goes much deeper.

WHY STORIES? *HOMO NARRANS*

Rudyard Kipling once said, "If history were taught in the form of stories, it would never be forgotten." The author of *The Jungle Book* and the youngest writer to win the Nobel Prize for Literature (1907) certainly knew a lot about telling memorable stories. But little did he know that in the ensuing century, a new branch of psychology would emerge that has amply confirmed Kipling's instinct about the power of stories.

One of the central tenets of *narrative theory* is that human thought is fundamentally structured around stories. People record and recall life experiences, as well as others' experiences, in the form of stories. Just think for a minute about the role of stories in daily life. We live in a world of stories. From children's fairy tales to Hollywood blockbusters, in all forms of media—books, television, radio, film, podcasts—stories are the currency of everyday life.

This has been true since or before the dawn of civilization. In oral cultures, for example, some reliable means was needed to transmit lore and information faithfully from generation to generation. All oral cultures that have been studied used (and still use) storytelling. Stories typically embed content into vivid imagery and characters that inspire our imagination and arouse our emotions. No doubt our ancestors discovered that knowledge embedded in story form was more memorable than in any other. It has been claimed, and reasonably so, that story is one of the most important human inventions. Indeed, we are such storytelling and story-seeking creatures that numerous experts have dubbed our species *Homo narrans* (the storytelling person).

One important thrust of modern research has been to understand why and how narrative plays such a crucial role in human thought. There are two universal aspects of story structure that combine to give it power. No matter who the storyteller, narratives are usually structured in ways that connect *cause and effect* (X happened, so Y happened). And stories are always composed of a sequence of events that are connected over time (first X happened, then Y, then Z). People thus use stories to *understand* cause and effect, and to connect series of events. Because narratives are constructed of causal links and some temporal order between events, stories carry the powers of both instruction and persuasion. People learn from stories because stories present a coherent argument in favor of some conclusion.

This fundamental, long-standing place of story in human culture has prompted scores of educators and psychologists to study the role of story in education and learning. They have found that narratives offer many advantages over formal explanatory text such as that found in textbooks. These benefits include increased recall, greater comprehension, and shorter reading times. For example, it is claimed that people remember information when it is woven into a narrative up to 22 times more than facts alone.

With respect to science content, narratives offer some very specific advantages over other kinds of texts. For instance, they are especially effective at conveying the scientific process. Jerome Bruner, one of the leading figures of narrative theory, noted:

> The process of science is narrative. It consists of spinning hypotheses about nature, testing them, correcting the hypotheses, and getting one's head straight. . . . The history of science . . . can be dramatically recounted as a set of almost heroic narratives in problem solving.

Bruner advocated more than 20 years ago that "our instruction in science from the start to the finish should be mindful of the lively processes of science making, rather than an account only of 'finished' science as represented in the textbook."

Even though the advantages of narrative are undisputed, as you well know, textbooks are much more of a fixture in science classes than are narrative stories. It is no wonder that legendary songwriter Paul Simon sang, "When I think back on all the crap I learned in high school, it's a wonder I can think at all."

And I quote that as a textbook author myself.

This state of affairs is more than a missed opportunity—it is a shame.

Without narrative to illuminate the making of science, there is little context as to what mystery or questions inspired an investigation or how a mystery was solved. And we get the wrong impression, or perhaps no impression at all, about who scientists are and what they do. Every biology student learns that Africa is the cradle of humanity, but there was so much more to the quest to solve one of the greatest mysteries of life than finding some old bones.

Mary Leakey's story involves all sorts of emotions—love, ambition, obsession, desire, conflict, frustration, disappointment, and, ultimately, the thrill of discovery. It is these feelings and other human qualities such as courage, persistence, sacrifice, and resilience that make for both good science and, if told well, a good and memorable story.

Which brings me to the rest of the stories in this book.

WHY THESE STORIES? *HOMO BIOLOGUS*

Good stories make for better and more memorable learning. But what makes a good story? Or more to the point, what makes a good science story—one that is worth precious time in the classroom, or at home?

Adam Gopnik, a brilliant writer for the *New Yorker* (and interestingly not a scientist himself), has made the most persuasive case I know about the essentials for good science stories. He argues that both good science stories and good scientific theories are *startling*; they *astonish* us with their claims. For example, he cites this story from physics: "Locked inside the nucleus of each little invisible atom is a force so vast it can destroy an entire city!"

Of course, biology has equally astonishing tales to tell: inside the nucleus of every cell are invisible molecules, within which the positions of atoms determine the characteristics of every living thing on earth!

Indeed, some claims are so astonishing they are initially disbelieved by scientists, and may remain disbelieved by nonscientists for decades, or centuries.

Gopnik notes that good science stories "startle us with their strangeness, but they intrigue us by their originality, and end up rewarding us with the truth, after an effort." But to be startling, to be astonishing, to be worth the effort, stories also have to address something fundamentally important—they have to answer a great question about nature. Great questions stem from great mysteries. And of all the phenomena on our planet and in the universe, life presents many mysteries that have inspired great questions, such as:

> Are species divinely created or the product of natural laws?
> Where did humans come from?
> How does like produce like?
> How does a single-celled egg become an adult?
> Why is the world green?
> What makes us sick?
> How can we prevent or cure a disease?
> How do our brains work?

These are some of the big questions tackled in the stories in this book. Professionals will recognize their answers as the foundations of modern evolutionary biology, genetics, microbiology, immunology, neuroscience, ecology, and so forth. But humans have contemplated such matters since long before these fields existed, indeed for millennia. All cultures have origin stories, and our hunter-gatherer ancestors had to be very observant about animals and plants, or we would not be here. The ancients grasped the basics of variation and reproduction as they transformed wild plants into food crops, and wild animals into livestock and companions. Similarly, untold numbers of "traditional" medicines were discovered long before the birth of the pharmaceutical industry, and vaccination was discovered and widely applied with no knowledge of the existence of viruses or the inner workings of the immune system. Indeed, one could make a pretty good case for another name for our species: *Homo biologus*.

What modern biology offers are new kinds of answers to some of the oldest and greatest questions humans have ever posed about life. These answers come in the form of startling discoveries and important truths about the workings of life, and of our place in nature.

My aim in this book is to present the stories of many of the seminal discoveries in biology, hence the title *The Story of Life*. The general plan of the book is to present classic stories along with more recent discoveries that offer new insights or apply new knowledge to medicine or managing nature.

I am keenly aware that every biology course entails a great deal of work, and more reading may be an unwelcome addition to that effort. I hope that burden will be lessened, however, as you discover much to be admired in these people and in their accomplishments. Most of these scientists paid mightily for their discoveries by enduring physical hardships or having their breakthroughs doubted or denied. Many bravely undertook great journeys or risky experiments. Given such great characters and high-stakes drama, I even dare hope that you just might enjoy reading these stories!

Or looking at it another way, as former Beatle George Harrison did in the preface to one of his writings, "I have suffered for this book, now it's your turn."

Sean B. Carroll
Chevy Chase, Maryland

THE STORY
of
LIFE

Great Discoveries in Biology

Part I

THE PROCESS OF SCIENCE

THE SCIENTIFIC PROCESS HAS proved to be the most powerful means humans have devised for learning about nature. One great strength of this process is that it is open to new information. New ideas are born continuously, and new evidence is constantly gathered and weighed against existing explanations. Sometimes long-established ideas are completely overturned. It is a process that, when followed, has the power to enable us to understand the often complex relationships between cause and effect in nature or in the human body. The first story in this section is all of this—a story about overturning established thinking with a revolutionary idea about the cause of a human disease. It is a shining example of the scientific process in action. But it is equally important to appreciate that when the scientific method is not followed or when evidence is swept aside, as in our second story, the consequences can be far-reaching, even tragic.

1

GUTS AND GLORY

Fortune favors the bold.
—LATIN PROVERB

OCTOBER 3, 2005, WAS a cool, cloudy Monday afternoon in Perth, Australia. Robin Warren and Barry Marshall, two longtime colleagues, were enjoying cold beers at their favorite pub on the Swan River. Warren's fish and chips had just arrived when his cell phone rang.

A man named Hans Jörnvall was calling from Sweden with some shocking news—Warren was to receive the Nobel Prize for medicine.

At first, Warren did not believe it was true—that the Nobel committee was calling *him*.

Once it sank in that the man on the other end was legitimate, Jörnvall explained to Warren that he had a big problem. The announcement was supposed to be made to the world in just half an hour, but he had been unable to reach Barry Marshall, with whom Warren was to share the prize. Jörnvall said he had called Marshall's home and his office but could not find him.

Warren laughed and said, "Oh, he is here with me at the pub! We are having a beer together." He passed his phone to Marshall.

At first glance, the two men were unlikely Nobel laureates. Warren was a pathologist and Marshall a gastroenterologist—they were not the kind of genetic pioneers that had garnered many of the recent prizes. Moreover, they worked in Perth, one of the most remote, albeit beautiful cities in the world,

2

far from the scientific and academic hubs of North America and Europe. But their discovery has changed the lives of millions of people, and rewritten medical textbooks.

SOMETHING UNUSUAL

The adventure began in 1979 on Robin Warren's 42nd birthday, not with a slice of cake, but with a slice of someone's stomach.

Warren was a consulting pathologist at the Royal Perth Hospital. One of his tasks was to examine biopsies from patients with various complaints. That day he was examining a biopsy from a patient with gastritis—an inflammation of the stomach. In his microscope, he noticed a thin blue line on the surface of the stomach lining. Under higher magnification, he thought he could see many small bacteria sticking to cells lining the stomach.

If true, his observation was very odd because Warren had been taught, and indeed it was medical dogma, that bacteria could not live in the normal stomach because of its acidity. Excited, Warren showed the specimen to a colleague, but he said he could not pick out the bacteria.

To assist his work in pathology, Warren had been experimenting with stains that helped to highlight features in tissue samples viewed under a microscope. He found that tiny bacteria were especially difficult to see against the background of human tissues. But he did discover one silver stain that marked certain bacteria well. He had no idea what sort of bacteria might be in his stomach biopsy, but he gave the stain a try.

It worked beautifully, revealing scores of small, curved, grayish bacteria against the yellow-brown background of stomach tissue (Figure 1.1). Warren thought the bacteria looked like those of a group named *Campylobacter* and said so in his formal pathology report. He added: "They appear to be actively growing and not a contaminant. I am not sure of the significance of these unusual findings, but further investigation of the patient's eating habits, gastrointestinal function and microbiology may be worthwhile."

His pathology colleagues believed that Warren had found bacteria, but it was unclear what the presence of the microbes meant. Were they just a one-time coincidence, or could they play a role in stomach disease? If the bacteria were significant, why had no else seen or reported them? One pathologist told him, "If you really believe they are important, see if you can find more cases."

Warren went looking for bacteria in biopsies taken from other gastritis patients, and found them often. But their presence could either be the primary

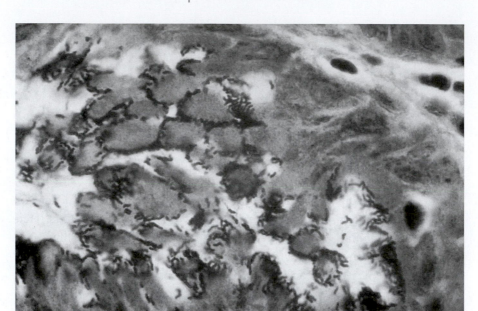

FIGURE 1.1 Robin Warren's First Case of Bacteria in a Stomach Biopsy Staining of stomach tissue from a patient with gastritis reveals many curved bacteria on the stomach lining. Photo courtesy of Robin Warren.

cause of the inflammation or result from a secondary infection of wounded tissue. To try to distinguish between these possibilities, he needed to examine a *negative control*—to look at healthy stomachs to see if bacteria were present. But that posed a problem—surgeons sought out diseased areas for biopsy, not normal-looking tissue. Eventually, Warren was able to review some normal-appearing biopsies and to determine that the bacteria were absent. This correlation was consistent with the possibility that the bacteria were causing disease, but Warren was still not able to convince others.

A COLLABORATOR APPEARS

He then met Barry Marshall, a young physician who was being trained in internal medicine at Royal Perth Hospital. Part of the training program involved rotations through different specialties. Marshall aimed to become a rheumatologist but began a stint in the gastroenterology division in July 1981. Marshall was hoping to do some research while he was there, and he was referred to Warren who was looking for a clinician to help him obtain some fresh biopsies to look for microbes.

Marshall visited Warren in his basement office, where Warren proceeded to show Marshall slide after slide of the strange, curved bacteria he had found in the inflamed stomach linings of gastritis patients. Marshall was intrigued because he also "knew" that the stomach was supposed to be sterile. He thought that identifying this unusual bacterium would make a good project, and agreed to work with Warren.

Marshall reviewed the case histories of the patients in whom Warren found bacteria. Their ailments were not limited to gastritis but also included ulcers, which are sores or breaks in the lining of the stomach or duodenum (the upper part of the small intestine). Marshall knew from clinical experience that ulcers were a very serious, usually lifelong condition that could cause debilitating pain. Ulcers were also associated with a much higher risk of stomach cancer. At the time, milder cases were treated with antacids and drugs that inhibited the production of stomach acid, but these treatments merely mitigated symptoms—they were not cures. More severe cases could prompt surgical removal of the lower third of the stomach.

In order to identify the bacterium, Marshall decided to try to culture it from fresh biopsies. As soon as the biopsies were removed, Marshall put them in a salt solution and delivered them to the hospital's microbiology department. The presence of the bacteria was confirmed by staining, and then a wash of the specimen was applied to bacteriological plates, placed in an incubator, and checked for bacterial growth 48 hours later.

Despite the evidence of bacteria in each biopsy, every attempt to culture them failed, sample after sample, week after week, for six months.

Marshall's stint in the gastroenterology rotation was ending, and he had to move on to a rotation in hematology. His project might have been a complete bust had he and Warren not also tried another way of testing the bacteria's significance. They figured that, if the bacteria were causing disease, then treatment with antibiotics might bring relief. They gave the antibiotic tetracycline to one elderly patient who had severe abdominal pain and whose biopsy revealed the presence of the curved bacteria. After two weeks, the man's pain and chronic gastric symptoms had disappeared.

The patient was ecstatic, but Warren and Marshall's peers were not swayed. This was just a single case, and Warren and Marshall still could not say whether the bacteria triggered the man's condition or were a secondary problem. Their colleagues urged the pair to undertake a large-scale, carefully controlled study to pin down the role of the bacteria in disease.

PROVING POSTULATES

The process for proving a particular microbe causes a specific disease has long been established in medical science. In the second half of the 19th century, Louis Pasteur in France and Robert Koch in Germany were leading proponents of the so-called germ theory of disease that asserted that certain diseases were caused by microorganisms. Koch proposed a series of criteria or "postulates" that needed to be fulfilled to prove a causal link between a microbe and disease, including these:

i. The infected tissue must show the presence of the microorganism not found in healthy tissue;
ii. The microorganism should be isolated and grown outside of the body;
iii. The microorganism should produce the disease when inoculated into a susceptible animal;
iv. The microorganism should be found in the area of induced disease.

Marshall, even though he had moved on to daily duties in the hematology ward, launched a 100-patient gastroenterology and microbiology study with several collaborators. The plan was to enroll patients undergoing stomach examinations and see whether the bacteria were present in sick patients and absent from normal stomachs. In the early morning before work and during his morning break, he would approach endoscopy patients and recruit them to take part in the study. Their biopsies were then kept on ice until Marshall was able to deliver them to the laboratory for microbiological examination and attempts to culture them.

A few months into the study, Marshall caught a very lucky break. On the Thursday before Easter 1982, a biopsy was taken from a man with an ulcer in his duodenum. Normally, the procedure was to culture the sample for 48 hours, after which the bacteriological plates were discarded. More than 30 cultures had turned up nothing. But on that Easter weekend the laboratory was short-handed, and the man's culture plates were not examined for five days, until the following Tuesday. This time, the technician saw small transparent colonies of an unfamiliar bacterium. The chief of microbiology called Marshall down to the lab to show him the exciting discovery growing on the plate.

Once they knew that the bacterium was slow-growing, they were quickly able to isolate it from another 11 patients. Electron microscopy revealed a spiral-like organism with multiple, tail-like flagella—similar to *Campylobacter*,

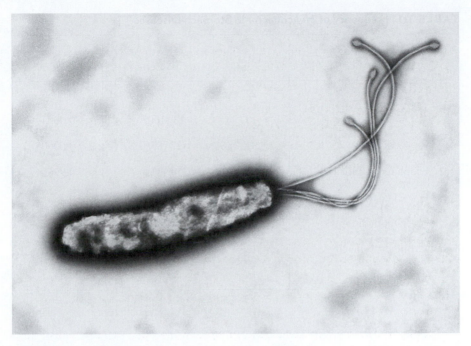

FIGURE 1.2 *Helicobacter pylori* The bacterium is about 0.5 μm in diameter and 2.5 μm long, with five flagella that have bulbous tips. Heather Davies/Science Source.

but different enough that it would eventually warrant its own name: *Helicobacter pylori* (Figure 1.2).

The ability to culture and recognize the organism was a key advance, but the main purpose of the 100-patient study was to see how the presence of the bacteria correlated with disease. The study concluded in May 1982, when Marshall was to relocate to a new posting at a hospital in Port Hedland, a mining town 1200 miles (2000 km) from Perth. Just before he left, Marshall spent a Saturday morning photocopying the clinical records of the 100 patients to take with him to his new assignment. Once in Port Hedland, he spent his evenings poring over the laboratory results and the patients' diagnoses, looking for any associations. He found that 65% of the patients had the bacteria, and nearly all of those had some evidence of gastritis. Moreover, all 13 patients with duodenal ulcer had the bacteria (Table 1.1). And just as importantly, people without the bacteria did not have ulcers (except for a few that were on painkillers known to cause stomach damage).

Statistical tests revealed that the relationship between bacteria and ulcers was very unlikely to be a random coincidence. It was time to share the results.

TABLE 1.1 Frequency of *Helicobacter* in Stomach Diseases

Endoscopic Appearance	Total	With Bacteria	p
Gastric ulcer	22	18 (77%)	0.0086
Duodenal ulcer	13	13 (100%)	0.00044
All ulcers	31	27 (87%)	0.00005
Esophagus abnormal	34	14 (41%)	0.996
Gastritis	42	23 (55%)	0.78
Duodenitis	17	9 (53%)	0.77
Bile in stomach	12	7 (58%)	0.62
Normal	16	8 (50%)	0.84

Data from Marshall and Warren (1984).

In February 1983, Marshall and Warren submitted an abstract of their findings to the Gastroenterological Society of Australia, which was holding its annual meeting in Perth. The abstract was rejected. The cover letter stated that the society was able to accept 56 of the 67 abstracts submitted—Marshall and Warren's report of the association between bacteria and stomach disease was rated near bottom of the heap (Figure 1.3).

Warren and Marshall had become accustomed to skepticism. The gastroenterology community had long believed that ulcers were a matter of lifestyle—including stress, diet, alcohol, and smoking—not due to an infection. The conventional thinking was that the appropriate treatment was to reduce stomach acid, and thanks to the prevalence of ulcers, those drugs were billion-dollar blockbusters. Marshall realized that gastroenterologists were not going to accept a revolutionary discovery based on 13 patients from Western Australia. He and Warren would have to do more to convince the medical community. The best way to do that was to publish their findings in prominent peer-reviewed journals, and to have others confirm them.

They reported their initial findings about the discovery and description of the bacteria in two papers in *The Lancet*, one of the world's oldest and most prestigious medical journals. In his report, Marshall speculated: "If these bacteria are truly associated with [gastritis] . . . , they may have a part to play in other poorly understood, gastritis associated diseases (i.e. peptic ulcer and

GASTROENTEROLOGICAL SOCIETY OF AUSTRALIA

145 Macquarie Street,
SYDNEY.   2000

Telephone 27 3288

17th March, 1983

Dear Dr. Marshall,

I regret that your research paper was not accepted for presentation on the programme of the Annual Scientific Meeting of the Gastroenterological Society of Australia to be held in Perth in May, 1983.

The number of abstracts we receive continues to increase and for this Meeting 67 were submitted and we were able to accept 56.

There were a large number of high quality abstracts which made it extremely difficult to choose those which should be accepted for presentation, and as you know, this is now done by a National Abstract Selection Committee which reviews the abstracts without knowledge of the Authors concerned.

The National Programme Committee would like to thank you for submitting your work, and would hope that this might be re-submitted in the future, perhaps following critical review from your colleagues.

My kindest regards,

Yours sincerely,

for Terry D. Bolin,
Honorary Secretary.

FIGURE 1.3 Even Nobel Prize–Winning Discoveries May Be Rejected at First Letter to Barry Marshall explaining that his abstract reporting the association of bacteria with stomach disease was not among the 56 selected for the 1983 Gastroenterological Society of Australia meeting. Letter courtesy of Barry Marshall.

gastric cancer).” The publications prompted other investigators around the world to search for the bacteria in their patients.

At this stage, however, Marshall and Warren had support for just two of Koch’s four postulates: the association of the bacteria with disease, and their ability to culture it. The remaining challenges were to show that the bacteria could induce disease in a laboratory animal, and that eradication of the bacteria eliminated disease. By 1984, the latter was going pretty well. Marshall was very excited to discover that the bacteria were extremely sensitive to bismuth (the main ingredient in Pepto-Bismol and long known to have antibacterial properties). He experimented with a combination of bismuth and the antibiotic metronidazole and was able to cure four out of four people.

Establishing the disease in an animal, however, was difficult. Marshall feared it was perhaps impossible. He had tried pigs because their large body size would allow Marshall to insert an endoscope into their stomachs, but inoculation of the pigs with bacteria was a complete failure. No disease formed and no bacteria were detectable in their stomachs.

Bacteria are often restricted in the species and tissues they can infect. If Marshall was going to convince the world that *Helicobacter* caused disease, he needed to find the right animal. Which species provided the best odds of getting infected?

Marshall realized the best candidate stared back at him in the mirror every morning.

WHEN ALL ELSE FAILS . . .

Marshall had been mulling testing the bacteria on himself for a while. His first concern, of course, was making himself sick with perhaps a chronic condition. But with his antibiotic regimen apparently working, he thought it might be reasonably safe for him to try.

The greater worry was scientific. He knew that if nothing happened in his experiment, if the bacteria did not colonize his stomach, if gastritis did not develop, then his hypothesis could be wrong. Or at the very least, the disease was more complicated than a bacterial infection.

He discussed the idea of swallowing the bacteria with a couple of colleagues, who seemed neutral about the idea. One microbiologist laughed and declined to “take the bug” himself. Clinical experiments on humans are regulated by institutional committees, but Marshall did not to apply for any formal approval, having decided that if he were turned down he would take the bacteria anyway—and probably lose his job.

He adopted a "don't ask, don't tell" strategy. He showed up one morning at the endoscopy clinic before it opened and asked a friend to put an endoscope into his stomach and take a biopsy. This would be Marshall's baseline sample before drinking the bacteria. It showed a normal, healthy stomach lining. That same day, Marshall biopsied a middle-aged man with gastritis. Laboratory tests revealed *Helicobacter pylori* that was sensitive to Marshall's antibiotic regimen. Marshall treated the patient for two weeks and confirmed that the infection was gone.

With a disease-causing germ in hand that he knew could be eradicated with antibiotics, Marshall decided the crucial test "was now or never." On the morning of his experiment, he skipped his breakfast and took a drug to reduce his stomach acid, figuring that might improve the ability of the bacterium to colonize his stomach. After fasting until 11 a.m., he was handed by a technician a beaker with about two teaspoons (10 milliliters) of broth in it, into which he had scraped a plate full of *Helicobacter* (about 1 billion bacteria). Marshall chugged the cloudy brown liquid in one gulp and, after waiting a couple of more hours, ate normally and went on with his work day. That evening, he told his wife, Adrienne, that he had started the experiment.

Marshall did not notice any symptoms for a few days, but then he woke up several mornings in a row nauseated, and vomited. His wife noticed his breath was "putrid," a fact his colleagues at work were too polite to tell him. After 10 days, Marshall again asked a colleague to take a biopsy. To his "joy," the bacteria were present in his inflamed stomach lining.

The experiment had worked. Koch's postulates had been met. *Helicobacter pylori* was a pathogen.

FROM HERETICS TO HEROES

Word of Marshall's experiment leaked out before he could document his experience. An American newspaper reporter woke Warren early one morning with questions about their work. When the reporter asked how Warren knew the bacteria caused disease, Warren told him about Marshall infecting himself. The story's headline was "Guinea-Pig Doctor Discovers New Cure for Ulcers . . . and the Cause."

Marshall wrote up his self-experiment for formal publication, referring to himself as a "volunteer" before disclosing that he was the subject described in the study. By this time in 1985, physicians in other parts of the world had begun confirming the presence of *Helicobacter* in their patients' stomachs— the link was growing stronger.

The medical community required still one more chunk of evidence to change its treatment of ulcers from focusing on stomach acids to treating them as an infection. That evidence came in the form of multiple clinical trials that tested the ability of various antibiotics to eradicate the bacteria from patients, and to prevent the relapse of ulcers. Trials carried out in Amsterdam, Vienna, Houston, and elsewhere all reported success in eradicating *Helicobacter pylori* in up to 90% of patients, and providing a lasting cure.

Antibiotics thus became and remain the mainstay of ulcer treatment today, and Marshall and Warren went from heretics to medical heroes. And on that joyous afternoon following their stunning call from Sweden, the two pioneers toasted their remarkable journey not with beer, or bacteria, but champagne (Figure 1.4).

FIGURE 1.4 Robin Warren and Barry Marshall Celebrate Their Nobel Prize

William Edwards/Reuters/Newscom.

END-OF-CHAPTER QUESTIONS

1. Map the steps involved in Warren and Marshall's discovery of the role of *H. pylori* in stomach disease, and its ultimate acceptance by the medical community, to the steps of the scientific process (see the Appendix).

   a. What observation triggered the first suspicion that bacteria might cause stomach disease?

   b. What logical questions did this observation prompt?

   c. Warren and Marshall proposed the hypothesis that bacteria caused stomach disease. What was an alternative hypothesis that could explain their initial observations?

   d. According to Koch's well-established postulates, if the bacteria caused disease, then what lines of evidence should bear that out?

   e. How did Marshall directly test whether bacteria could cause gastritis?

   f. What predictions did the bacterial theory of stomach disease make about treatment?

   g. Why was the bacterial cause of stomach ulcers ultimately accepted?

2

FIRST DO NO HARM

I will respect the hard-won scientific gains of those physicians in whose steps I walk.
—MODERN HIPPOCRATIC OATH

IN FEBRUARY 1998, ANOTHER report in the medical journal *The Lancet* sparked a revolution—more like a revolt—of a different kind than Warren and Marshall's (Chapter 1).

The paper described twelve young patients with gastrointestinal disorders who also exhibited developmental problems characteristic of "regressive autism." The term described children who first develop normally, then exhibit losses in acquired language and social skills. This was in contrast to "classical" autism in which acquisition of such skills never occurs or is severely delayed.

Because of the apparent delayed onset of the symptoms, parents and physicians suspected that there was some kind of environmental trigger for the condition. A team of thirteen authors associated with the Royal Free Hospital and School of Medicine in London reported that parents or physicians of eight of the twelve children linked the onset of symptoms to vaccination with the measles, mumps, and rubella (MMR) vaccine. In the eight children, behavioral symptoms were reported to occur an average of 6.3 days after vaccination (range of 1–14 days). Five children were also reported to have adverse reactions to the vaccine (rash, fever, delirium; and in three cases, convulsions).

The authors stated that they had not proved a link between MMR vaccination and the gastrointestinal/behavioral syndrome. They noted, "If there is a causal link between measles, mumps, and rubella vaccine and this syndrome, a rising incidence might be anticipated after the introduction of this vaccine." And indeed, the rate of autism appeared to be rising. The authors urged, "Further investigations are needed to examine this syndrome and its possible relation to this vaccine."

At the time, U.K. government policy was for every child to receive the first of two MMR shots between the ages of 13 and 15 months. The possibility that this policy could actually be causing harm would be shocking. A press conference was called at the Royal Free Hospital on the day of publication. The dean of the medical school urged that the results on such a small number of children be interpreted cautiously. He was worried that public confidence in the MMR vaccine could be damaged. "Measles vaccines are among the safest and most effective vaccines ever developed," he said.

However, the lead author of the study, Dr. Andrew Wakefield, thought action was necessary: "One more case of this is too many. It's a moral issue for me and I can't support the continued use of these three vaccines given in combination until this issue has been resolved."

The genie was out of the bottle.

FIRESTORM

The British press ran stories with headlines such as "Alert over Child Jabs" (jab is British slang for vaccine injection) and "Ban Three-in-One Jab, Urge Doctors."

The possible link between a routine vaccine and a rising incidence of autism understandably worried parents, physicians, and policy makers around the globe. The MMR vaccine had been in use in the United States for 30 years, much longer than in the United Kingdom. In April 2000, the U.S. Congress held a hearing to get to the bottom of the matter. Representative Dan Burton of Indiana, the chairman of the House Committee on Oversight and Government Reform, had been touched personally by the issue. "My grandson . . . , after receiving nine shots in one day, quit speaking, ran around banging his head against the wall, screaming and hollering and waving his hands, and became a totally different child. And we found out that he was autistic." Burton continued, "He was born healthy. . . . Then he had those shots and our lives changed and his life changed."

Several parents of autistic children then testified, expressing their conviction that the MMR vaccine had triggered their child's syndrome. Blame was placed on the drug companies that manufactured the vaccines. As one parent put it, "They manufacture products that are required for every child in this country. While seeking greater profits, [they] have lost sight of the medical community's original goal—to protect children from harm."

Andrew Wakefield also testified. He revealed that he had now studied (but not published results for) an additional 150 children, and stated that the vaccine had caused autism in all but 4. Several more scientists, some collaborators of Wakefield's, supported Wakefield's hypothesis of the vaccine causing autism.

That viewpoint, however, was not unanimous. Brent Taylor, also a professor at the Royal Free, disagreed. Taylor had examined 293 cases of autism and found no difference in the age of children diagnosed with autism who had received the vaccine and those who were never vaccinated. Taylor also found no association between the timing of vaccination and the onset of symptoms. And he also found no evidence for an increase in the rate of autism after the introduction of the MMR vaccine in England. Taylor had published his findings the previous summer in *The Lancet*.

"The belief that MMR is the cause of autism is a false hope," Taylor testified. "There is no evidence that immunizations are involved."

Despite Taylor's testimony against a link, press coverage of the hearing spread awareness of the possible connection between the MMR vaccine and autism. Andrew Wakefield was in the center of the media spotlight, cast as David against the Goliath of the medical establishment. Wakefield appeared on the most widely watched American news show, *60 Minutes*, and stated that he would not give his own children the combination MMR vaccine. He was also featured on the British Broadcasting Company's Sunday evening program *Panorama* in an episode entitled "MMR: Every Parent's Choice." Each program presented stories of parents who witnessed the heartbreaking decline of their children. Certainly, something was to blame, and the MMR vaccinations—or vaccines in general—seemed to be logical suspects. Most medical experts who were interviewed on camera disagreed, and tried to underscore the safety and effectiveness of vaccines.

The medical establishment was criticized publicly and often. One editorial in the *Telegraph* was entitled "Shame on Officials Who Say MMR Is Safe." At a rally in Washington, D.C., Wakefield spoke:

> We are in the midst of an international epidemic. Those responsible for investigating and dealing with the epidemic have failed. Among the reasons for this failure is the fact they are faced with the prospect that they themselves may be responsible for the epidemic. . . .

I believe that public health officials know there is a problem; they are, however, willing to deny the problem and accept the loss of an unknown number
of children on the basis that the success of public health policy—mandatory
vaccination—by necessity involves sacrifice.

In the face of such controversy, the decision to vaccinate or not posed a
dilemma with potentially high stakes. Whom or what were parents to believe?

At the same rally, Wakefield outlined the path forward, "our quest for truth
through science—a science that is compassionate, uncompromising, and
uncompromised."

That quest would entail more studies, require more scrutiny, and uncover
some very disturbing surprises.

THE QUEST FOR TRUTH

Wakefield had testified that he had extended his findings of a link between
vaccination, gastrointestinal disease, and autism to many more children.
However, a series of independent studies by researchers in the United Kingdom, Canada, and Finland did not detect any association between the MMR
vaccine and autism. Several months after the BBC's MMR program aired, a
very large study of Danish children was published in the *New England Journal of Medicine*. Comparison of autism rates between a group of ~440,000
children who had received the MMR vaccine and ~97,000 who had not been
vaccinated revealed no difference in the risk of autism. Nor did the study find
any association between the age at the time of vaccination, or the time since
vaccination, and the diagnosis of autism.

These failures to confirm a link between MMR vaccine and autism did not
quell the controversy. The timing of the battery of childhood immunizations
and parents noticing behavioral symptoms in their children seemed too close
to be explained as mere coincidence. So, if the culprit were not MMR itself,
what else could it be?

The next suspect was a chemical that was present in several vaccines. Thimerosal, a mercury-containing compound with antibacterial activity, had been
used for decades as a preservative, although never in the MMR vaccine.
Although the amount of thimerosal present in vaccines was very small, mercury and various forms of mercury were known to be toxic to humans and
other animals. As a precautionary measure, thimerosal was removed from
childhood vaccines in the United States in 2001. But this action, taken at a time

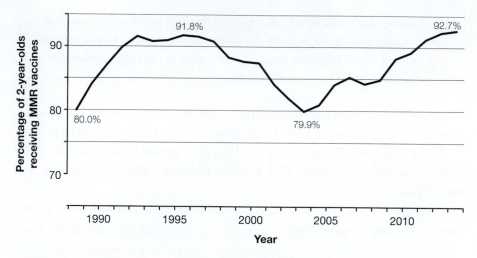

FIGURE 2.1 MMR Vaccination Coverage in England 1988–2014 The percentage of 24-month-old toddlers who received the MMR vaccine is plotted. The Wakefield study appeared in 1998. Data from U.K. National Health Service Immunisation Statistics, 2013–2014.

of much debate about the safety of vaccines, appeared only to inspire new worry.

Magazine articles and television programs sounded the alarm, often misreporting the concentration of mercury in vaccines. Politicians again joined the chorus of concern, publicly doubting the government agencies responsible for vaccine recommendations and safety. By 2004, vaccination rates in Great Britain had declined from about 92% of children before the Wakefield report to about 79% afterward (Figure 2.1). That drop worried public health experts because the larger the fraction of unimmunized children, the greater the risk of infections spreading through the population.

Was thimerosal, and particularly the cumulative dose of thimerosal from multiple vaccines, a cause of autism? Epidemiologists once again examined data on vaccine use and autism to search for any link. But again, large-scale studies in Finland and the United States found no association between exposure to thimerosal-containing vaccines and the incidence of autism.

The repeated failure to detect a causal link between vaccination and autism in large, controlled studies in many countries reassured many people of the safety of vaccines. But, that failure raised the question of why couldn't any other scientists repeat Wakefield's findings? Some critics had pointed out that the subjects in the initial study were not a random sample of children; they had been selected because of their combination of symptoms, and the proximity of their onset to MMR vaccination. Even if the children were a random group, statistically speaking, a study is more likely to detect a spurious association in such a small sample of twelve. Perhaps the contradiction between larger studies and Wakefield's were a matter of study design, or of small numbers?

No, the explanation was worse, much worse.

LIES AND MORE LIES

On February 18, 2004, six years after the publication of the initial report in *The Lancet*, Brian Deer, a journalist for London's *Sunday Times*, went to the journal's offices to meet with the editor in chief and several senior staff. Over the course of five hours, he shared what he had learned in four months of investigation into the research and researchers involved in the original study published in 1998. His revelations would eventually unravel not only the vaccine–autism story, but several careers.

Deer's first bombshell was that funding for Wakefield's research (about $90,000) had been arranged by a law firm that was preparing a product liability lawsuit against vaccine manufacturers. This fact raised the question of a conflict of interest: Had Wakefield's research and the study's findings been tainted by the aims of the lawsuit and the source of Wakefield's funding?

The second stunner was that several of the children in the study were those of clients of the law firm who were to be plaintiffs in the lawsuit. The affected children did not just happen to appear at the Royal Free Hospital, but were referred to Wakefield as part of building a case for a lawsuit. This fact raised the issue of ethical medical practice: Were the children being studied to understand the cause of their symptoms, or to prepare a lawsuit? The third disturbing issue was that neither of these facts were disclosed in *The Lancet* article, nor to most of Wakefield's coauthors.

The editor of *The Lancet* was shocked. Several of Wakefield's coauthors were angry.

Two weeks after Deer published his story in the *Sunday Times*, 10 coauthors, but not Wakefield himself, issued a "retraction" of the interpretation raised in the paper of a possible causal link between the MMR vaccine and autism.

Journalist Deer subsequently discovered that beyond the research funding, Wakefield had personally received a much larger amount in fees (~$750,000) from the same law firm over a period of many years. The possibility that

Wakefield's study had been tainted by undisclosed conflicts of interest and medical misconduct, and the impact the purported vaccine–autism link had on public health, prompted the British medical licensing board (the General Medical Council) to launch an official investigation.

The inquiry took two and a half years and called 36 witnesses. In 2010, the General Medical Council found that Wakefield's conduct had been "dishonest," "irresponsible," and "misleading" and that he had acted "contrary to his duties as a medical practitioner" and with "callous disregard" toward the children he had studied. The board found Wakefield guilty of serious professional misconduct and ordered him stripped of his right to practice medicine.

Further investigation by Deer revealed that most, if not all, of the children's cases in the original article were reported inaccurately or misrepresented to such a degree that the *British Medical Journal* declared the study not just scientifically and ethically flawed, but "an elaborate fraud." *The Lancet* officially retracted the entire original article.

CONSEQUENCES

Although no causal link to autism was confirmed, the increased concern over vaccine safety and the subsequent decline in vaccination rates had effects on public health. Before the introduction of a measles vaccine in 1968, Great Britain frequently experienced measles epidemics that struck as many as half a million people. Once most of the population was vaccinated, measles became relatively rare, and those cases originated outside of the country. But in 2007–2008, a decade after the Wakefield article, 2347 cases were reported, as many in the previous 11 years combined. And measles has become reestablished in Britain. Child vaccination rates have since rebounded to above 90%, but the disease remains a challenge due to large numbers of teenagers who were not vaccinated as children during the autism controversy.

In the United States, vaccination has long been mandatory for school-age children. However, exemptions may be given for medical reasons as well as for religious or "philosophical" reasons in most states. The rate of nonmedical exemptions increased over the period of 2000–2011 to about 2% nationally, but several individual states have exemption rates greater than 5%. Moreover, there are geographic pockets of communities with much higher rates of unvaccinated people. Measles was declared eliminated in the United States in 2000, when the only cases of the disease were imported from foreign visitors. Since that time, however, numerous outbreaks of the disease have occurred, largely among unvaccinated individuals and groups. In 2014, 667 people in 27 states were infected, the largest number since 2000.

The threat of the importation of measles, and further outbreaks, remains high because an estimated 20 million cases occur per year worldwide, and the disease is very contagious. Measles remains one of the leading causes of death among young children. Almost 115,000 children died in 2014. That is a frightening statistic, but it is one-fifth of the number of deaths that occurred in 2000, thanks to efforts that have immunized over 80% of children around the globe.

A second important consequence of the vaccine–autism controversy is that it drew attention, energy, and resources away from investigations into the actual cause of the syndrome. There has long been evidence of a strong genetic influence on the occurrence of autism, with males, and twins and siblings of autistic children, having much greater incidence of autism. In recent years, several gene mutations associated with autism have been identified that provide important leads into understanding the genetic and neurological basis of the syndrome.

Despite the massive epidemiological evidence to the contrary, and advances in understanding the role of genes in autism, the vaccine–autism link has been difficult to vanquish in the public mind. A 2015 Gallup poll of American adults revealed that 6% believed that vaccines cause autism, 42% believed they don't, and 52% were unsure.

END-OF-CHAPTER QUESTIONS

1. (a) Considering the steps in the scientific process summarized in the Appendix, how did the scientific process initially struggle in the vaccine–autism controversy? (b) How did the scientific process succeed?
2. For discussion. In your view, what appears to be the strongest evidence against a link between vaccination and autism?
3. The following table presents the incidence of autism and related disorders in Danish children who received or did not receive the MMR vaccine (1991–1998).

Vaccination	Number of Children	Autistic Disorder	Other Autism Spectrum Disorder
Total	537,303	316	422
Unvaccinated	96,638	53	77
Vaccinated	440,665	263	345

   a. Calculate the frequency of autistic disorder among (i) all children in the study; (ii) the unvaccinated children; and (iii) the vaccinated children.

   b. Compare the frequency of autistic disorder among the unvaccinated and vaccinated children. Do the results support the hypothesis of a link between MMR vaccination and autism?

4. The R in the MMR vaccine stands for rubella, a disease caused by the rubella virus. While rubella is a short-term and mild infection in many people, the virus can cause congenital rubella syndrome (CRS) in infants born to infected mothers. CRS symptoms include cataracts, deafness, heart defects, and intellectual disabilities.

   In 1964–65, before a vaccine was available, a rubella epidemic swept the United States, affecting approximately 12.5 million people. During the outbreak, 11,000 pregnant women lost their babies, 2100 newborns died, and 20,000 babies were born with CRS. Today, fewer than 10 cases of rubella occur in the United States per year, all of which result from exposure outside the country. Offer one argument for and one argument against continuing rubella vaccination in the United States.

Part II

HEREDITY

PEOPLE HAVE UNDERSTOOD THE basic phenomenon of inheritance for millennia, long before scientists learned anything about any mechanisms. The domestication of plants and animals began approximately 10,000 years ago when our ancestors began selectively breeding variants of wild plants and animals to establish crops and livestock. They clearly understood that variants of traits could be passed on to the next generations through breeding. It was not until the 19th century and the pioneering experiments of Augustinian monk Gregor Mendel (1822–1884) with pea plants that anyone started to understand how traits were passed from parents to offspring. Although published in 1866, at a time when there was great interest in the new Darwin–Wallace theory of evolution and in inheritance, Mendel's work and ideas went largely unnoticed. The modest priest often told his friends, "My time will come."

Mendel's time did come, although like many pioneers in biology, including several in this section, he did not live to see it. Mendel's work was "rediscovered" around 1900. Over the next half century, the mystery of inheritance was finally cracked with the discoveries of the chromosomal basis of heredity (Chapter 3), of DNA as the chemical basis of inheritance (Chapter 4), and of the structure of DNA (Chapter 5) and how genetic information is encoded, inherited, and altered by mutation. These fundamental discoveries gave birth to a technological revolution in genetic engineering, and created a new era in medicine with human diseases now being treated successfully with gene therapy (Chapter 6).

3

LIKE BEADS ON A NECKLACE

The future of the world lies in the hands of those who are able to carry the interpretation of nature a step further than their predecessors.
—THOMAS H. HUXLEY

LIKE MANY STUDENTS TODAY, Thomas Hunt Morgan attended college in his hometown. But it is hard to imagine any school today that would try to enforce the rules that Morgan was obliged to follow at the State College of Kentucky in Lexington in 1880. The three hundred all-male students and ten faculty were governed by 189 rules set forth by the board of trustees and strictly enforced by the school's president. For example, possessing any book besides a textbook, or newspapers or magazines, required special permission from the president, as did leaving one's room during study hours. And woe to any students who discussed college authorities, positively or negatively—that was strictly forbidden.

Despite the strict rules, and a few demerits for "disorder in the hall" and "tardiness at chapel," Morgan excelled at the school. In his senior year, however, Morgan found himself nearly failing his French class, despite having worked hard. The young man who would later write some of the major rules of heredity was almost a victim of his own pedigree.

Morgan's was a prominent Kentucky family with deep American roots. His ancestors on his father's side arrived in 1636, while his mother was the grand-daughter of a colonel in the Revolutionary Army and later governor of Maryland, as well as the granddaughter of Francis Scott Key, who composed "The

Star-Spangled Banner." It was his father's brother, John Hunt Morgan, whose exploits came back to haunt Tom (as he was called).

When the Civil War (1861–1865) broke out, John Hunt Morgan joined the Confederate cause and rose to become a brigadier general. In the summer of 1863 (the same summer of the Gettysburg Campaign and battle), General Morgan led a bold raid of more than 2000 cavalrymen into the northern states of Indiana and Ohio. Morgan's Raiders, as they became known, took horses and supplies from Union civilians, and many prisoners, too, as they skirmished with state militias and battled Union troops in their 1000-mile ride over 46 days. Unfortunately for Tom, one of the prisoners taken by his uncle was his future French teacher, Professor François Helveti, who was forced to ride a mule 90 miles from Cincinnati to Lexington while facing backward. More than 20 years later, Helveti was not feeling any forgiveness toward the Morgan family.

Tom ultimately passed Helveti's class. And he would eventually lead his own charge into new territory and earn the Morgan name new fame—in science.

PROFESSOR MORGAN

Ever since he was a young boy, Tom loved to roam the beautiful Kentucky countryside, to collect various creatures, and to scour railway cuts through hillsides for fossils, often taking groups of other boys along with him. When he was about 10, he was given two rooms in the attic to store and display his collections of birds, butterflies, eggs, and other treasures. After graduation, Morgan pursued his keen interest in animal life and its history by enrolling at Johns Hopkins University for graduate school in zoology.

Each of the deepest mysteries in biology—heredity, development, and evolution—appeared wide open as Morgan began his studies in 1886. The Darwin–Wallace theory of evolution had been published decades earlier, in 1858–1859 (Chapter 7), and the fact that life had changed over time was well accepted, but there was still great uncertainty and skepticism about the mechanism of evolution. In particular, there was no explanation of how changes were inherited and passed on between generations. Mendel's experiments on inheritance in pea plants had been published 20 years earlier (1866) but had gone completely unnoticed. And the profound mystery of development, of how a single fertilized egg cell gave rise to a new individual, had not been penetrated at all.

Like many biologists of his day, Morgan was interested in all three processes because they had to be connected in some way. Changes in animal form over time had to be due to changes in development, and such changes had to be inherited. But as a practical matter, he had to settle on one field to start. He chose development. After graduate school, he became a professor at Bryn Mawr College, where he worked for many years on all sorts of marine animals and their embryos. He loved to devise experiments without using any fancy apparatus. For example, to test the effect of gravity on the development of sea urchins, he put their eggs in test tubes, hooked them up to a bicycle wheel that was powered by a water-driven motor, and spun them around.

Morgan's specific interest in heredity was rekindled by a developmental puzzle that humans had pondered for ages, namely, what determined the sex of a fertilized egg? Morgan and other biologists believed that external factors such as nutrition or temperature influenced which sex developed. In 1903, Morgan wrote, "It may be a futile attempt to try to discover any one influence that has a deciding influence for all kinds of eggs." But very soon after Morgan wrote those words, one of his own Bryn Mawr students launched a study that would shed crucial light on the puzzle of sex.

Nettie Maria Stevens was an unusual scientist in many respects. First, she was female, a rarity because the opportunities available to women in science in the 1890s were very limited. Bryn Mawr was in fact the first college to grant PhDs to women. Second, when she began her doctoral work at age 39, after a distinguished career as a teacher, she was much older than most students. And third, she was exceptionally talented. By the time she completed her PhD in 1903, she had published nine papers, and earned awards that allowed her to conduct part of her studies at leading institutions in Italy and Germany. Morgan was very enthusiastic in recommending Stevens for a special fellowship to continue her research. He wrote, "Of the graduate students that I have had during the last twelve years I have had no one that was as capable and independent in research work as Miss Stevens. . . . Miss Stevens has not only the training but she also has the natural talent that is I believe much harder to find."

Stevens wanted to study the number and behavior of chromosomes in various insects, and their possible relationship to male and female sex. By the early 1900s, it was known that cells contained visible structures dubbed *chromosomes*, but their biological function was unclear. Microscopic studies of plants and animals had revealed that the number of chromosomes differed among species. Inspection of some species with particularly large cells revealed

(A)   (B)

FIGURE 3.1 The Heterochromosome of the Darkling Beetle (A) Nettie Stevens at her microscope. (B) Left, nine large chromosomes and one small chromosome visible in developing sperm of the darkling beetle; right, ten large chromosomes visible in another developing sperm. (A) Photo courtesy of Bryn Mawr College Library; (B) Stevens (1905), Plate VI, images 186 and 187.

individual chromosomes that were different in size, and therefore not identical. Moreover, it had been shown that chromosomes were duplicated prior to cell division and that each daughter cell received one copy of each duplicate. The individuality and behavior of chromosomes prompted the hypothesis that chromosomes were in some way responsible for heredity.

Still, no trait had been linked to a specific chromosome, nor had any trait been traced from the chromosomes of a parent to its offspring. Stevens selected five insects for study and closely examined the chromosomes of adults, eggs, and sperm.

In a beetle known as the common mealworm or darkling beetle (*Tenebrio molitor*), she counted twenty large chromosomes in adult females, but nineteen large chromosomes and one small chromosome in males. In the unfertilized eggs, she always found ten large chromosomes, but she discovered two types of sperm in the species: one type had ten large chromosomes and the other had nine large chromosomes and one distinct small chromosome (a "heterochromosome") (Figure 3.1).

To Stevens, the situation in *Tenebrio* appeared to be a clear case of sex being determined by a difference in one chromosome. She proposed that sperm carrying the smaller heterochromosome determined the male sex and those that contained the ten large chromosomes determined the female sex. The egg contributed its ten chromosomes to each sex.

In this beetle at least, sex seemed to have a simple basis. But how the differences in chromosomes influenced sex was unclear. Stevens speculated that sex could be a matter of the quantity or quality of chromosomal material. She could not draw any generality about sex determination, however, because none of the other species she studied possessed a similar heterochromosome. More data from more species would be needed.

Over the next several years, Stevens expanded her survey to 50 species of beetles, and found the heterochromosome in 12. Others also found similar heterochromosomes in a subset of insects. But to Stevens and Morgan, as well as to other biologists, the inconsistent finding of a heterochromosome in males was perplexing. Clarity about the mechanism of sex determination, indeed the mechanism by which all traits are inherited, would come from the study of one small insect in particular.

THE WHITE-EYED FLY

In 1900, three botanists independently rediscovered Gregor Mendel's work and helped make it known to the scientific world. The Augustinian monk had conducted years of breeding experiments with common garden pea plants that overturned one crucial, commonly held notion about inheritance. It was widely believed that mating between two different forms (e.g., tall and short) would produce offspring with intermediate characteristics because of "blending inheritance." But Mendel discovered that certain characteristics did not blend, rather they were "dominant" or "recessive" to one another. For example, crossing of tall and dwarf plants, or those with smooth or wrinkled seeds, yielded all tall plants or all smooth seeds, respectively, in the first hybrid generation.

Furthermore, Mendel discovered that when such hybrids were crossed, they produced tall and dwarf plants, or smooth and wrinkled seeds, respectively, in a 3:1 ratio. Thus, the original maternal or paternal characteristic reappeared, and was not blended. The observation that traits were inherited as discrete entities and not blended led Mendel to propose that traits were inherited as "particles" whereby offspring receive one particle from each parent, although Mendel could say nothing about the nature of such particles.

Mendel also found that when hybrids for two traits were crossed, the traits were inherited separately such that all combinations of tall/dwarf plants and smooth/wrinkled seeds appeared in a ratio of 9:3:3:1. From his observations, Mendel articulated two "laws" of heredity: (i) the law of segregation, which states that during gamete formation, two copies of each hereditary "factor" segregate such that offspring receive one factor from each parent; and (ii) the law of independent assortment, which states that for two or more characteristics, the hereditary factors governing each assort independently during gamete formation and therefore can be inherited in every combination in the offspring.

Mendel did not use the terms *genetics* or *genes*—the latter was not coined until 1909. Morgan's contemporaries described *factors* that determined the appearance of traits. But what were these particles and factors?

Chromosomes were obvious candidates for Mendel's particles, but there were few chromosomes relative to the potentially very large numbers of traits and factors. And the association between chromosomes and sex was intriguing but inconsistent. In moths and birds, for example, it was the female that had a heterochromosome. The situation was so confusing that Morgan could not see how chromosomes could determine sex, and he even at times doubted whether Mendel was correct.

There had to be a functional mechanism to heredity, and Morgan decided to look for it. After moving to Columbia University in 1904, he launched a variety of studies with all sorts of animals, including rats, mice, and insects, in the hopes of obtaining mutations like Mendel's.

Morgan tried to induce mutations in various insects by injecting all kinds of substances—acids, alkali, sugars, salts—into the parts of pupae where reproductive cells formed. Nothing worked. He tried other insects such as the fruit fly *Drosophila melanogaster*, even exposing their larvae to radioactive radium. For more than two years, he had no luck. Nothing unusual emerged in his bottles. "Two years wasted," he told one visitor to his laboratory. "I have been breeding those flies for all that time and I've got nothing out of it."

Then, in early 1910, he spotted a single male fruit fly with white eyes, as opposed to the normally vivid red eyes of both sexes. Morgan carefully bred the white-eyed male with a red-eyed female and patiently waited for the offspring to emerge more than 10 days later. Virtually all had red eyes.

Morgan then bred siblings of this red-eyed first generation (designated F_1) with one another to produce a second (F_2) generation. The white-eyed trait reappeared in a 1:3 ratio to the red-eyed trait. This was exactly how a recessive Mendelian trait should behave, except . . . all of the white-eyed flies were

FIGURE 3.2 The White-Eyed Mutant Fly Left, a white-eyed male *D. melanogaster* similar to the fly Morgan discovered in 1910. Right, a normal male fly with red eyes. Courtesy of Steve Paddock.

male (Figure 3.2)! Morgan was ecstatic. It appeared that the white-eyed trait was inherited along with sex.

Morgan tried to reconcile his results with what was known about chromosomes in *D. melanogaster*. Nettie Stevens had discovered that *D. melanogaster* had four chromosomes, and that males carried a small heterochromosome. One possibility was that the white-eyed trait was associated with the heterochromosome. However, when Morgan bred F_1 females from a cross between white-eyed males and red-eyed females with white-eyed males, approximately half the males were white-eyed and half the females were also white-eyed. Morgan could only explain white-eyed females by supposing that the *white* factor was not associated with the Y chromosome, but with the X chromosome. White-eyed females (XX) had two copies of the *white* factor, whereas white-eyed males (XY) had one copy that manifested itself because the Y chromosome lacked any eye color factor.

A close association between a factor and a chromosome was exciting, although not enough for Morgan to draw any great generality. Morgan stopped short of concluding that the *white* factor was part of the X chromosome. He needed more evidence, and more mutants.

Whatever the earlier problem was in finding mutants vanished. After his discovery of the white-eyed fly, this first trickle of mutant flies turned into a flood. Morgan identified another sex-linked mutation for yellow body color and one for miniaturized wings. One or two new mutants a month joined his growing menagerie. Morgan was so busy with his flies that he confessed in a paper in the prestigious journal *Science*, "These mutations have appeared in such rapid succession that my time has been almost entirely consumed in producing pure strains of the new forms, which can be utilized later for a thorough study of . . . inheritance."

He would soon get some timely help with his fly farm.

THE FLY ROOM

In the fall of 1909, for the only time in his 24-year career at Columbia, Morgan stepped in for a colleague and gave the opening lectures in the introductory zoology class. Those lectures would change several lives and the history of genetics, although not because of Morgan's teaching ability—he was an average lecturer at best. It was because two undergraduates, Alfred Sturtevant and Calvin Bridges, just happened to be taking the class. Each would later seek out Morgan and become his close collaborators.

Sturtevant grew up on a farm in Alabama and had traced out the pedigrees of his father's racehorses as a hobby. In his junior year, he brought Morgan a paper he was working on about the inheritance of horse coat colors. Morgan was so impressed, he helped Sturtevant get the paper published, and offered him a place in his lab working with fruit flies. Bridges was academically a year behind Sturtevant, so Morgan started him out washing food bottles in the lab. That lasted only until Bridges spotted a fly with a different eye color, a new mutant called vermilion eye, through the thick glass of a milk bottle. Bridges promptly got a desk alongside Sturtevant and Morgan in the Fly Room, as Morgan's lab would become known.

The small room had eight desks crammed into it, with a large kitchen table in the center where the fly food was prepared. The workers examined flies at their desks using a hand lens (microscopes were not yet standard equipment), counted and scored their appearance, selected those to be mated, and placed them in half-pint milk bottles that Morgan "borrowed" from the university cafeteria. Morgan first used mashed, fermented banana as fly food, which resulted in an overwhelming stench that disgusted neighbors throughout the building. The lab later switched to agar containing banana juice. A piece of paper was inserted into each bottle to absorb excess moisture and give the flies a drier perch. Always frugal, Morgan often used envelopes from his correspondence (sometimes unopened).

(A)   (B)

FIGURE 3.3 The Fly Room (A) Calvin Bridges and Alfred Sturtevant in the Fly Room. (B) Thomas "The Boss" Morgan in the Fly Room. Morgan was so camera shy, Sturtevant had to hide a camera and take the picture remotely using a string.

(A) Courtesy of the Archives, California Institute of Technology; (B) courtesy of the Archives, California Institute of Technology.

Despite the great difference in age (Morgan was 44, and Sturtevant and Bridges were 19 and 21 when they joined the lab), Morgan established a close camaraderie with his young team. Unlike in the European system, where professors were usually addressed formally and seldom challenged, Morgan encouraged open discussion and the free flow of opinions. It did not matter who came up with a new idea or criticism; the group all chimed in. Morgan also encouraged discussion outside of the lab by holding meetings at his house on Friday evenings where science papers were discussed over beer and crackers. Morgan was very frugal with institutional funds, but he was generous with his own pocket, and quietly helped many students financially. They referred to him as "The Boss" out of both affection and respect (Figure 3.3).

One of the major topics of discussion was the physical relationship between the different sex-linked factors that had been found. If the factors resided *on* the X chromosome, then they should segregate together. But in crosses of flies that carried both the white and rudimentary wing factors, sometimes the two traits did not segregate together in the offspring.

It was Morgan who came up with an explanation for the behavior of the two traits. He had read a paper by F. A. Janssens, who studied the formation of amphibian germ cells. Under the microscope, Janssens saw that pairs of chromosomes intertwined with each other during meiosis. He suggested that the chromosomes actually broke and exchanged parts. Morgan realized that this "crossing-over" (as it was later known) would allow for factors that were not close together to be exchanged between the two X chromosomes.

Sturtevant was sitting in Morgan's office as the Boss explained his reasoning. The college senior suddenly realized that if the strength of linkage varied with the distance between factors, then it might be possible to determine the order of and distance between factors on a chromosome based on how often traits segregated together or apart. He went right home and, neglecting his other homework, spent most of the night calculating the frequency of crossing-over between the sex-linked factors. Based on the frequency with which they did not cosegregate, the yellow body color factor appeared to be located very close to the white eye color factor, but relatively far from vermilion eye color and very far from the wing factors. In one night, Sturtevant had constructed the very first genetic map.

Morgan now had very strong evidence that factors resided at different places on chromosomes. In just one page in the journal *Science*, he formulated a crucial addition to Mendel's laws. Mendel had emphasized the independent assortment of genetic factors in his second law. Morgan sought to account for the cosegregation or "coupling" of some sex chromosome factors he observed in *Drosophila*. Morgan wrote, "We find coupling in certain characters, and little or no coupling in other characters; the difference depending on the linear distance apart of the chromosomal materials that represent the factors." He explained, "The results are a simple mechanical result of the location of the materials in the chromosomes, and of the method of union of homologous chromosomes, and . . . the relative location of the factors in the chromosomes."

Morgan envisioned the factors (soon to be called "genes") on the chromosome like beads on a necklace, such that when the two chromosomes/necklaces paired, they could break and exchange sets of genes/beads (Figure 3.4).

Sturtevant, Bridges, and Morgan soon discovered that other genes exhibited no coupling with the X chromosome genes, but that other groups coupled with one another. For example, black body, curved wing, and vestigial wing coupled with one another and five other factors, but not with four other genes that were themselves coupled together. The team had evidence for three "linkage groups" in *Drosophila*: one group on the X chromosome, and two more groups that were not on the X. They concluded that each linkage group resided on a different chromosome.

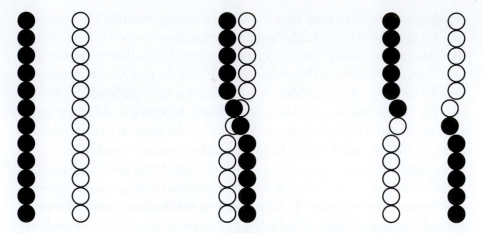

FIGURE 3.4 Genes Are like Beads on a Necklace The diagram illustrates crossing-over between a pair of homologous chromosomes (left). Genes are depicted as white and black circles. During crossing-over (center), the chromosomes are broken and exchange parts, such that groups of formerly linked genes do not segregate with one another (right). Morgan (1915), p. 426.

THE CHROMOSOMAL THEORY OF HEREDITY

A handful of papers on *Drosophila* and the convictions of the Fly Room gang were not sufficient to persuade the world that they had solved the mysteries of chromosomes and heredity. Morgan, true to the team spirit of the Fly Room, enlisted his young accomplices—Sturtevant, Bridges, and another student, Hermann Muller—to be coauthors of a book that would review all of their evidence and the entire field of heredity, to be entitled *The Mechanism of Mendelian Heredity* (1915).

The book was a watershed in biology. The mysteries of heredity that just a few years earlier had appeared so murky to everyone, including Morgan, now had a concrete physical explanation and strong experimental support. One reviewer said, "The chromosome theory begins to appear as one of the great miracles of human achievement."

Tragically, Nettie Stevens did not live to see the chromosomal theory take hold. She died in 1912 of breast cancer. Morgan wrote a long tribute to his former student in *Science*: "Miss Stevens had a share in a discovery of importance, and her name will be remembered for this. . . . Her single-mindedness and devotion, combined with keen powers of observation; her thoughtfulness and patience, united to a well-balanced judgment, accounts, in part for her remarkable accomplishment."

After receiving their bachelor's degrees, Sturtevant, Bridges, and Muller stayed in the Fly Room to obtain their doctorates under Morgan. Indeed, the first two men remained 17 years at Columbia, continuing to work independently on many facets of heredity alongside the Boss, who was awarded the 1933 Nobel Prize in Physiology or Medicine.

END-OF-CHAPTER QUESTIONS

1. What observations led Mendel to conclude that traits were inherited as particles from each parent?
2. Why was it difficult to figure out whether chromosomes determined sex?
3. Discuss the role the following pieces of evidence played in the making of the chromosomal theory of heredity:
   a. The discovery of the white-eyed fly.
   b. The microscopic evidence for chromosome pairs crossing over.
   c. The coupling of factors on the *Drosophila* X chromosome.
4. On accepting his Nobel Prize, Morgan said, "Personally, I realize of course that the work in genetics has not been accomplished by any one individual or group of individuals, but has been world wide and the outcome of many hands and minds." What does the making of the chromosomal theory of heredity suggest about the importance of collaboration in science?

4

WHO WOULD HAVE GUESSED IT?

The greatest single achievement of nature to date was surely the invention of the molecule DNA.

—LEWIS THOMAS, *THE MEDUSA AND THE SNAIL*

THE PLANES WERE COMING again.

After a respite of more than a month, Londoners were once more taking cover at the wail of air raid sirens, and listening anxiously for the drone of approaching aircraft. More than 60 times since September 7, 1940, the first night of the Blitz when over 250 German bombers pounded the capital with high explosives, bombs, and incendiaries, residents had braced themselves against the din of antiaircraft guns, the whistling of falling bombs, and the stomach-churning roar of explosions and collapsing buildings.

In the first eight months of Germany's campaign to bomb Britain into submission, more than 20,000 civilians were killed in London alone. Hundreds of thousands of buildings were destroyed or damaged, and more than 1 million people lost their homes. Countless landmarks, including Buckingham Palace and the Houses of Parliament, were badly damaged. Shipping and transportation were crippled—food and fuel were scarce and rationed.

But somehow London and Britain had endured. In a national radio address, Prime Minister Winston Churchill praised the British people for standing up against Hitler's "blackmail by murder and terrorism." He lauded the police, fire, ambulance, and rescue workers who had responded to the country's ordeal with tremendous bravery. Churchill also praised the medical and public health authorities who had managed to prevent any increase in illness, despite the crowded conditions in air raid shelters and the disruptions to water and sewage lines.

One of those authorities was Dr. Frederick Griffith, a 62-year-old micro-biologist who helped set up the Emergency Public Health Laboratory at the outset of the war. A staunch patriot, Griffith also contributed his aluminum pots and pans to help build Spitfires, bought war bonds, and steadfastly refused to move out of London "for any German."

With all citizens doing their part, Churchill reassured his listeners, "We shall not fail or falter; we shall not weaken or tire."

But this night, April 16, 1941, would be one of London's worse nights and greatest tests. The attack would be much larger, one of the heaviest since the war began. Beginning around nine o'clock in the evening and lasting until nearly dawn, 685 bombers would discharge their payloads over the city. In addition to the 890 tons of high explosives and 150,000 incendiaries, the Germans dropped parachute mines, large naval mines attached to parachutes that would descend slowly and land on building rooftops, then detonate seconds later. The weapon had a much larger blast area than bombs that detonated on impact, and had proven particularly effective at maximizing destruction in cities.

Just after midnight, a parachute mine landed on Griffith's house at 75 Eccleston Square in the historic and handsome borough of Westminster. The ensuing blast leveled the building and damaged several nearby structures. Ambulances and rescue parties were sent to the area. They worked through the night and into the morning to search the rubble for survivors and victims. The frantic scene was repeated at hundreds of locations across the city.

Griffith did not survive; he was one of more than 1000 Londoners who perished in the raid. A soft-spoken, reclusive man who was unmarried, had no children, and few close friends, Griffith was not the sort of personality to be long remembered by posterity. But years earlier, he had made an astonishing discovery that would lead to the surprising revelation that DNA was the chemical basis of heredity, and ensure that his legacy in the story of genetics would endure.

ADVENTURES WITH PNEUMONIA

Griffith had devoted his life to studying the bacteria that caused diseases, including tuberculosis, pneumonia, and scarlet fever. He firmly believed that progress in treating infectious diseases and preventing epidemics would require precise knowledge of the microbes responsible. One of the first requirements was to be able to identify and classify different species of bacteria, or different strains of one species involved in infections.

Much of Griffith's focus had been shaped by war. At the end of his life, he was studying bacteria isolated from war wounds. Twenty years earlier, following the end of the First World War, he focused on bacteria that played a part in a deadly worldwide pandemic. Toward the end of that war (1918), an influenza pandemic erupted that eventually killed 20 million to 50 million people, many more than died in combat. For most victims, it was not the flu virus itself that was lethal, but secondary infections that followed. One of the most common culprits was the bacterium *Streptococcus pneumoniae*, a normal inhabitant of nasal and lung linings. Griffith was keen to understand how and why this "ubiquitous and apparently harmless organism may suddenly become more pathogenic . . . and propagate an epidemic."

It was known that there were several types of *S. pneumoniae*. Types I, II, and III were each distinguished by antibodies from people or animals that had been exposed to the bacteria and reacted specifically with chemical structures on the gelatinous capsule that coated the bacteria. A fourth heterogeneous group was called Type or Group IV. Griffith discovered that *S. pneumoniae* were even more diverse than these four types. When cultured on a bacterial plate, colonies of the four major types have a smooth, rounded, and glistening appearance due to their capsules. Griffith discovered another variant in which colonies were small with irregular, rough edges, and lacked capsules (Figure 4.1). When bacteria were injected into mice, Griffith observed that all of the "smooth" or S forms were virulent (caused disease) and fatal, whereas the "rough" or R form was not ("avirulent").

From mice inoculated with samples from one advanced infection, however, Griffith obtained an unexpected mixture of both smooth and rough colonies of bacteria that grew up on an agar plate. When reinoculated into mice, three colonies of the rough forms were avirulent, as expected, but one rough colony caused a fatal infection. This was the first time Griffith or anyone else had observed a virulent rough strain. The rough colony belonged to Type IV.

Griffith decided to investigate this unusual microbe. Since previous rough forms were derived from cultures of smooth bacteria, he wanted to see whether he might be able to revert the rough form back into a fully smooth form. After a long series of passages in mice and culture in the laboratory, he obtained a small shiny colony similar to the smooth type that was still virulent. He then tried to repeat the feat with a rough, avirulent Type II strain. Griffith found

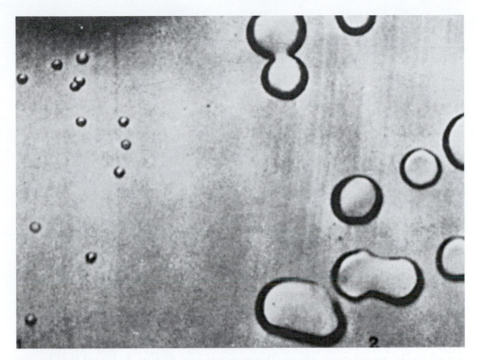

FIGURE 4.1 Smooth and Rough Forms of *S. pneumoniae* The smooth, virulent forms appear so because of a polysaccharide capsule around the bacterium; rough forms lack the capsule and form smaller colonies. Image courtesy of Rockefeller Archive Center.

that by starting with a large dose of bacteria in mice, he was able to obtain a smooth, virulent form.

How this change in type occurred was a mystery. The latter result gave Griffith the notion that something about the large dose of bacteria was important for transforming a strain from the rough avirulent form to a smooth virulent form. The structures on the capsular surface of the bacteria that determined their types were known to be complex carbohydrates. Rough bacteria lacked a capsule and those carbohydrates, but Griffith wondered whether there might be some trace of a carbohydrate-synthesizing enzyme that remained, such that when a large number of rough cells was congregated, some might have their type-specific carbohydrate restored.

Thinking that such an enzyme might be vulnerable to heat, he started testing whether bacteria killed by heating at various temperatures might also be able to transform a small quantity of live avirulent bacteria into virulent bacteria. The experiment worked—the combination of killed bacteria and live, avirulent rough bacteria injected into mice produced smooth, virulent colonies. Control experiments demonstrated that neither preparation alone (killed bacteria or live avirulent bacteria) was sufficient to produce any transformation.

In his initial experiments with killed cells, Griffith transformed a rough strain with killed preparations of the same type. He expected that this striking effect would be limited to combinations of cells of the same type. Griffith then tried mixing rough avirulent strains derived from one type with killed preparations of another type and injected the combination into mice. To his surprise, he obtained smooth, virulent strains—all of which were of the killed bacteria type!

The result was so unexpected that Griffith was careful to be absolutely certain that the killed preparation contained no live bacteria that would undermine the experiment. It did not. Griffith repeated the experiment with various combinations of killed and live types (Figure 4.2). The rough strains were transformed to the killed bacteria type.

Griffith wasn't the only one who was surprised. His report of the transformations of *S. pneumoniae* types stunned fellow microbiologists. Oswald Avery, for example, an eminent American *Streptococcus* expert, believed that bacterial types were stable. He found it hard to accept that *Streptococcus* could change from one type into another. That was until someone in his own lab repeated Griffith's results while Avery was away on summer vacation. Griffith had earned the reputation for being very careful, and he described his experimental methods in such exact detail that other laboratories were also able to promptly confirm his findings.

But just what was the substance responsible for the dramatic transformation of types? What sort of chemical could induce a heritable change in bacteria? The identity of the transforming substance was still unknown at the time of Griffith's death 13 years after his discovery. Griffith himself did not pursue the question—he was more concerned with the medical and epidemiological significance of his findings. But across the Atlantic, far from the nightmare in London, one group had picked up the trail, and was getting close to an answer that would stun the scientific world.

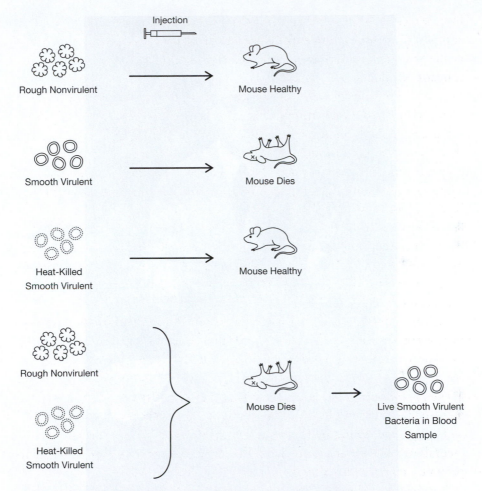

FIGURE 4.2 Griffith's Discovery of the Transformation of Bacterial Types Schematic of Griffith's experiment in which he injected rough or smooth strains, a heat-killed smooth strain, or a mixture of a live rough and a heat-killed smooth strain. The unexpected result was the lethality of the mixture, and his recovery of a live virulent, smooth strain of the same type as the killed strain.

FESS

Oswald Avery and Fred Griffith had two important traits in common—both were trained physicians with a passion for understanding *Streptococcus* pneumonia, and both had a very cautious scientific nature. Each man was determined to understand and to help thwart the disease, and careful not to overstate

FIGURE 4.3 Oswald T. "Fess" Avery

Science History Images/Alamy Stock Photo.

any implications of their own work. But in other ways, the two men were opposites. Whereas the painfully shy Griffith avoided social situations, Avery was welcoming to any visitor or stranger, and far more outgoing and comfortable in conversation. His warmth attracted talented junior scientists to work in his group, and he was gifted with such impressive verbal skills that students and his colleagues at Rockefeller University Hospital in New York affectionately referred to him as the Professor or Fess (Figure 4.3).

Avery initially had strong doubts about the transformation of *Streptococcus* types, but once he was convinced the phenomenon was real, he started to chase after the transforming substance. The experiments were very difficult, and there were other, more urgent priorities, so the work progressed in fits and starts, with some long dormant spells. In October 1940, Avery and Colin MacLeod, a young Canadian physician, together resumed the hunt.

Over the previous decade, Avery's group had made two important advances. First, they had succeeded in transforming bacteria in laboratory culture, thereby eliminating the time-consuming process and variable outcomes of inoculating live mice. And second, they were able to produce an extract from bacteria that replicated the transforming activity of killed bacteria. With a simpler test for activity, and an active extract, they could potentially purify and identify the transforming substance (or what Avery called the transforming "principle").

The thick, syrupy extracts contained all of the classes of large molecules in bacterial cells: proteins, lipids, carbohydrates, and nucleic acids. That complexity was not the only challenge to purification—the transforming activity itself was destroyed by enzymes present in the extracts. In 1940, there were very few established techniques for purifying or analyzing large molecules. One good rule of thumb for any purification strategy was to be sure to start with as much active material as possible. Despite efforts to make extracts in the same way, however, their activity was highly variable, and sometimes zero. "Disappointment is my daily bread," Avery would often say.

MacLeod and Avery focused on trying a number of techniques to separate the extracts into different classes of molecules. What unfolded was essentially a process of elimination in which different classes of molecules were removed one by one from the extracts. In the winter of 1940–41, MacLeod and Avery learned that the chemical chloroform ($CHCl_3$) was a very effective way of removing protein (and destructive enzymes) from the extract. They also found out that adding ribonuclease enzyme to destroy ribonucleic acid (RNA) had no effect on activity. Once they figured out how to obtain more active material from bacteria, they scaled up their preparations from 2- or 3-liter cultures of cells to 50-liter batches.

One of the separation steps involved precipitation of the transforming activity with alcohol. This precipitate contained a large amount of the capsular polysaccharide, which raised the possibility that it played some role in transformation. MacLeod had to leave the lab for other duties and Maclyn McCarty, another young physician, joined Avery's lab. McCarty assessed the role of the polysaccharide by destroying it with a specific enzyme. He found that the transforming activity was unaffected. The transforming activity was made of something else.

McCarty was able to make extracts that were essentially devoid of proteins or polysaccharides. What substances remained in the extract? In chemical tests, the extract was positive for the sugar deoxyribose. This was an indicator of the presence of deoxyribonucleic acid (DNA), but it did not prove that the transforming principle was made of DNA because other substances were also present. McCarty noticed that DNA had the same fibrous and viscous appearance as his preparations of transforming principle. So McCarty decided to test the ability of various crude enzyme preparations that broke down (depolymerized) DNA to affect transforming activity. He found a perfect correlation between the depolymerization of DNA and the destruction of transforming activity.

Still, these results were only suggestive that DNA *could* be the substance responsible. McCarty and Avery knew the stakes were too high to get it wrong. They had to eliminate the possibility that some other contaminating substance carried the activity. What they needed was as pure a preparation of bacterial DNA as they could get. McCarty figured out how to fractionate and purify DNA and he found that essentially all of the transforming activity resided in his DNA fraction.

Up to this time, biologists knew nothing about the function of DNA, not even whether it was present in most species. And chemists knew only that DNA was composed of just four nucleotides—its structure was unknown. Proteins, made of long chains of 20 different amino acids, were believed to be responsible for specific biological activities within cells.

In May 1943, Avery wrote to his brother Roy, also a scientist, to share the news:

> For the past two years, first with MacLeod and now with Dr. McCarty I have been trying to find out what is the chemical nature of the substance which induces this specific change. . . . Some job, full of headaches and heartbreaks. But at last <u>perhaps</u> we have it. . . .
>
> In short, this substance is highly reactive and on elementary analysis conforms <u>very</u> closely to the theoretical values of pure desoxyribose nucleic acid. . . . (Who could have guessed it)

The work had not yet been submitted for publication or reviewed, so Avery maintained a mixture of caution and excitement:

> If we are right and of course that is not yet proven, then it means that . . . by means of a known chemical substance it is possible to induce <u>predictable and hereditary</u> changes in cells. This is something that has long been the dream of geneticists.

NO NOBEL?

The discovery was published in February 1944, when the world and much of the scientific community was engrossed in the war. It took a while for biologists to become aware of the report. Reactions were mixed. Some found it difficult to believe that DNA, which was thought to be a polymer made up of four repeating bases (ACGTACGTACGT) could carry specific information. Some thought the transforming activity might be a phenomenon peculiar to bacteria. Others, however, thought it was the most exciting breakthrough in all of biology.

It would take several years and further discoveries by others to convince the skeptics, and to shed light on just how DNA could carry information. Avery, who turned 67 in 1944, had put off retiring to pursue the transforming substance. He fully retired in 1948 and died in 1955, two years after Watson and Crick deciphered the structure of DNA (Chapter 5).

One might think that the discovery of the function of DNA would certainly have earned Avery the Nobel Prize. But Nobels are not awarded posthumously, and the Nobel Committee did not recognize the significance of Avery's work in time. The chairman of the committee later admitted that overlooking Avery was one of the Nobel's greatest oversights. Of course, Avery's discovery had only been made possible by Griffith's original, difficult, and surprising experiments. Griffith, too, was ineligible, and also never recognized.

END-OF-CHAPTER QUESTIONS

1. What was Fred Griffith's very surprising observation that eventually led to the discovery of the transforming principle and DNA as the hereditary material?
2. What early advances enabled Avery's group to pursue the eventual purification of the transforming principle?
3. The following table is from Avery, MacLeod, and McCarty's paper on the transforming principle. It shows which crude enzyme preparations contained certain activities and which were able to inactivate the transforming principle.

Inactivation of Transforming Principle by Crude Enzyme Preparations

Crude Enzyme Preparation	ENZYMATIC ACTIVITY			
	Phosphatase	Tributyrin Esterase	Depolymerase for DNA	Inactivation of Transforming Principle
Dog intestinal mucosa..........	+	+	+	+
Rabbit bone phosphatase....	+	+	−	−
Swine kidney phosphatase	+	−	−	−
Pneumococcus autolysates......	−	+	+	+
Normal dog and rabbit serum..........	+	+	+	+

Avery et al. (1944). Republished with permission of Rockefeller University Press, conveyed through Copyright Clearance Center, Inc.

    a. Is this evidence that the transforming principle is DNA?

    b. Is this proof that the transforming principle is DNA?

4. Oswald Avery and Fred Griffith both died before the significance of their discoveries was realized. Is it fair that Nobel Prizes are not awarded post-humously? What might be an argument against the awarding of Nobel Prizes after death? An argument for their award?

5

THE SECRET OF LIFE

It is a capital mistake to theorize before one has data.
Insensibly one begins to twist facts to suit theories, instead of
theories to suit facts.

—SHERLOCK HOLMES IN *A SCANDAL IN BOHEMIA* (1891)
BY ARTHUR CONAN DOYLE

THE PROCESS OF SCIENCE is often compared to a detective story, where the solution to some mystery is arrived at by the dogged pursuit of evidence and clever deductions by a scientist (the lead detective). In the case of the mystery of the structure of DNA, and how it helped explain the mechanism of inheritance, the detective story unfolded more like a game of *Clue*.

In the popular board game, first introduced by Parker Brothers in 1949, just before the case of DNA was cracked, the goal is to be the first of several players to solve a murder by correctly identifying which character committed the crime, with what weapon, and in which room of a mansion. Each player assumes the identity of one of the six characters in the game and moves that game piece around the board making "suggestions" of the solution, for example, "I suggest the crime was committed by Colonel Mustard with the candlestick in the library." Players then try to prove the suggestion is false by secretly showing the accuser a clue card that proves the suspect, weapon, or room is not in the special solution envelope. Once a player is confident that they know the solution, they make an "accusation" to win the game.

The game of DNA likewise involved six main players, several false suggestions, a fair amount of sharing—and not sharing—of important clues, and some accusations. The cast of characters involved was every bit as quirky as the fictional Miss Scarlet, Professor Plum, and Mr. Green. They included:

Erwin Chargaff, erudite, intense, often critical of others, professor of biochemistry at Columbia University, College of Physicians and Surgeons, New York;

Linus Pauling, brilliant, outgoing, politically outspoken, and famous professor of chemistry at Caltech (Pasadena) who in 1951 deciphered the major structures formed by proteins;

Maurice Wilkins, a shy and reserved Englishman, who worked as a physicist on the Manhattan Project (the atomic bomb) in the United States, and turned to biology after the war. Wilkins was a member of the biophysics unit at King's College London;

Rosalind Franklin, confident, sometimes impatient with others, and an expert in using X-rays to decipher the structure of molecules. Franklin joined Wilkins' unit at King's College in 1951.

Each of these players would hold vital clues, and might have solved the mystery on their own, or had an even better chance of success by working together. But none of the four worked together. When Pauling and Chargaff first met, Pauling took an instant dislike to Chargaff and avoided him. Wilkins on the other hand did believe that he and Franklin were to work together, but intense personality conflicts drove them apart. The structure was ultimately deciphered by perhaps the unlikeliest detectives, two men with barely any credentials who had even been barred at one point from pursuing the case—Dr. Watson and Mr. Crick:

James Watson, brash, ambitious, who received his PhD in zoology at age 22 in 1950. He went to Europe seeking the training he would need to crack DNA;

Francis Crick, talkative, adept at theory and mathematics, who first studied physics then switched to biology after the war. He was still a graduate student (age 35) at the University of Cambridge in 1951.

One might think that there would be more than six people on the trail of something as important as the structure of DNA (Figure 5.1). But that importance is largely a matter of hindsight. No one could have known just how much the structure itself would reveal about the mechanisms of heredity. Moreover, even several years after Oswald Avery's discovery of the transforming principle (Chapter 4), there were still lingering reservations and uncertainty

Watson Crick Wilkins Franklin Chargaff Pauling

FIGURE 5.1 The Players

Central Press/Getty Images; Express/Getty Images; Bettmann Archive/Getty Images; Donaldson Collection/Michael Ochs Archives/Getty Images; Horst Tappe/Getty Images; New York Public Library/Science Source.

about DNA being the chemical basis of heredity. Chargaff was convinced— indeed, he was so persuaded by Avery's studies that he shifted his laboratory from working on fat and proteins to working on DNA. But Pauling was not so sure. He thought that proteins with their great diversity of structures and functions should somehow be involved in heredity. Furthermore, most scientists who were keenly interested in genes and heredity did not have the training or expertise to tackle structural biochemistry.

Watson himself was in this latter group.

GETTING INTO THE GAME

No one was more convinced of the importance of DNA than Watson. He was certain that a Nobel Prize awaited whoever could crack the structure. But Watson knew little chemistry. He had avoided the subject as an undergraduate at the University of Chicago (1943–1947), where he was much more interested in birds and dreamed of someday becoming the curator of ornithology at the American Museum of Natural History in New York. He did learn, however, about Avery's work in a genetics class, and he was spurred to read the short book *What Is Life?* (1944) by Erwin Schrödinger.

The eminent physicist (Nobel Prize 1933) probed the question of the difference between nonliving matter and life and focused in particular on the mystery of heredity. Writing before Avery's discovery, Schrödinger inferred that the three-dimensional arrangement of atoms in some polymer had to explain the two main properties of heredity. Those properties were the stability of life, such that traits were passed faithfully from generation to generation; and the mutability of life, the ability of life to evolve over time. Such great powers begged a physical explanation.

Watson was driven to "finding out the secret of the gene." Rejected from graduate school at both Harvard and Caltech, Watson enrolled (at age 19) at

Indiana University where he did learn a lot of genetics, but again very little chemistry. After a solid but unremarkable PhD project, Watson's mentor suggested that the best place for him to get the training he wanted in the biochemistry of nucleic acids was in Europe, specifically in Denmark. Watson sailed for Copenhagen in September 1950.

The first and most important thing Watson learned in Copenhagen was that he was in the wrong place, or at least that he had arrived at the wrong time, to learn what he needed. His mentor, distracted by an impending divorce, left Watson to fend for himself. After several unproductive months, his advisor suggested that Watson accompany him that spring to the zoological station in Naples. He spent most of his time enjoying the sun, walking the streets, or reading articles.

Watson was not the only scientist seeking a little time in the sun. Maurice Wilkins also went to Naples that spring to attend a meeting on the structure of biomolecules, courtesy of his boss in London who could not make the trip himself. Watson sat in on the meeting, trying to make sense of the technical talks. He heard nothing revealing or inspiring, until Wilkins took the podium and talked about his and his student Raymond Gosling's efforts to make X-ray diffraction images of DNA.

In this technique, a narrow X-ray beam is focused on a molecular crystal or fiber that contains a regular array of atoms. The atoms diffract the X-ray in different directions depending on their identity and position within the molecule. Analysis of the resulting diffraction pattern on photographic film can reveal the detailed structure of the molecule. Wilkins had discovered that purified DNA could be drawn out into very thin threadlike fibers containing many molecules. Near the end of his talk, Wilkins showed a photograph of an X-ray diffraction pattern of one such fiber of DNA that seemed to contain a lot of detail.

Watson was thrilled. Before that moment, Watson had feared that DNA might be an irregular structure, very difficult or even impossible to decipher. The features in the picture suggested that DNA had a regular, repeating structure that could be solved. It had never occurred to him or his mentors that he needed to learn about X-ray crystallography. To get into the game, he had to get out of Copenhagen and go to where the action was.

Watson evaluated his options. He tried to approach Wilkins in Naples about coming to London, but Wilkins slipped away before Watson could put the question to him. Shortly after he returned to Copenhagen, a spectacular flurry of papers appeared on the structures of proteins by Linus Pauling. Caltech would certainly be one place to learn crystallography, but Watson doubted that Pauling would be willing to waste his time on someone with so little training. The other hub of structural work was in Cambridge, England. Watson's graduate advisor made the introductions and the eager crystallographer-to-be happily said goodbye to Denmark.

CAMBRIDGE

Watson joined the laboratory of Max Perutz, one of the leaders of the Cambridge group, where he met an assortment of researchers working on the structures of biomolecules, primarily proteins. Among them was Francis Crick, a 35-year-old graduate student. Crick's first attempt at a PhD—in physics—was derailed by the war when he was recruited into the Admiralty and spent seven years designing magnetic and acoustic mines. His abandoned PhD project became a direct casualty of the war when the building housing the apparatus he had constructed was destroyed by a direct hit from a parachute mine. It was a mercy killing. Crick thought his project—to determine the viscosity of water under different conditions—the "dullest problem imaginable."

After the war, Crick was also drawn toward biology by reading Schrödinger's little book. He was persuaded by the argument that just like nonliving matter, life must also operate by the laws of physics and chemistry. He hoped that his training in physics would give him some foundation for studying the molecules of life. Crick went to Cambridge in 1947 but had still not earned his PhD when he met the much younger Watson (aged 23) in October of 1951.

The two men hit it off right away. Both loved to talk, and talk they did— at the ritual tea and coffee breaks, at lunch in the nearby Eagle pub, whenever they were together, often several hours a day. In Crick, Watson found a partner who not only believed that DNA was more important than proteins, but also could tutor him in the interpretation of X-ray crystallographic data. In Watson, Crick had a partner who was up-to-date on the latest findings in genetics. Their animated conversations did not go unnoticed. Crick had a booming laugh and spoke faster and louder than anyone else around the laboratory or the building. The pair was soon assigned their own office.

Even though Crick's thesis work was on proteins, he and Watson quickly decided that they should tackle the structure of DNA together. They agreed that they could not solve it by staring at X-ray pictures alone. They were impressed by Pauling's approach to protein structure, which involved building Tinkertoy-like models and using the rules of chemistry to figure out which atoms sit next to each other. They decided they would take that approach as well. They also resolved to search for the simplest structure that was consistent with whatever data were available. They hoped that it might be some kind of helix, as Pauling had found in proteins. There was no sense in worrying about much more complicated structures until they ruled out simpler ones.

A few important details were known about the chemical formula of DNA. Just two years earlier, chemists at Cambridge had determined that the polymer was made up of a backbone consisting of alternating phosphate and sugar (deoxyribose) groups, to which the nitrogen-containing bases adenine, guanine, cytosine, and thymine were attached. The polymer was a long chain of smaller units known as nucleotides, consisting of a base attached to a sugar and phosphate group (Figure 5.2).

Watson and Crick assumed that the nucleotides occurred in an irregular order. If the order were always the same, then all DNA molecules would be identical and could not carry gene-specific or species-specific information. But despite that irregular order, DNA had to form some regular structure as indicated by X-ray pictures.

Among the many unknowns was the number of chains. Wilkins told Crick that the diameter of the DNA molecule was thicker than it would be if made of only one chain. There might be two, three, or even four chains. In any case, some sort of chemical bonds had to hold the chains together. It was unknown whether those might be ionic bonds involving the negatively charged phosphate groups, or weaker hydrogen bonds. Nor was it known whether the bases or sugar–phosphate backbones were inside or outside the chains.

There were a lot of possibilities. To narrow them down, the pair needed access to good X-ray pictures. Rather than try to make their own images, which could take many months, if they succeeded at all, they decided to reach out to Wilkins. There were potential priority and turf issues in asking to see Wilkins' data. Crick and Wilkins had been friends for several years, however, and Wilkins readily agreed to make the short train ride from London up to Cambridge to chat.

Wilkins shared his suspicion that DNA was made of three chains, but he did not share Watson and Crick's enthusiasm for building models. Wilkins thought that more and better pictures of DNA were needed. The trouble was, he confided to Watson and Crick, he and Rosalind Franklin, who had joined the unit early that year, were not getting along at all. The two scientists had opposite personalities. Wilkins was quiet, had the habit of looking away from people as he spoke, and shunned conflict. Franklin was direct, sometimes brusque, and she enjoyed the give-and-take of debate. After sizing up Wilkins,

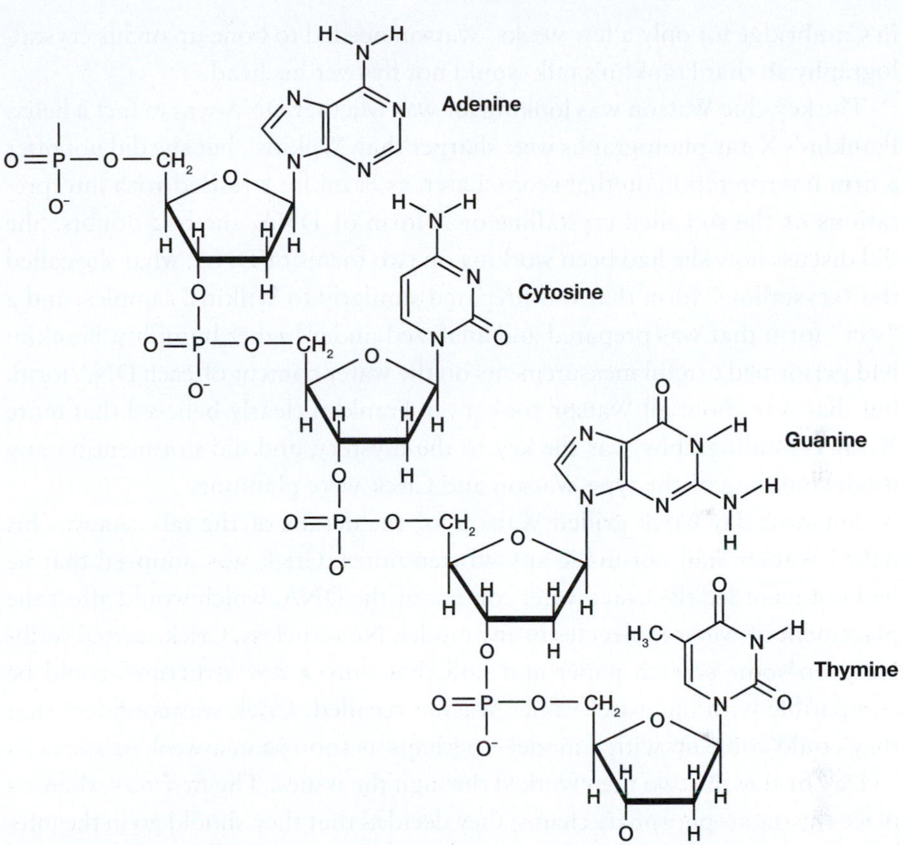

FIGURE 5.2 DNA Is Composed of Nucleotides The backbone is made up of alternating sugar (2-deoxyribose) and phosphate groups. Each sugar has attached to it a side group consisting of either a purine base (adenine or guanine) or a pyrimidine base (cytosine or thymine).

Franklin rejected the notion that she and Wilkins would work together, although Wilkins had given her all of his good crystalline DNA. Franklin had taken over supervision of Raymond Gosling, and told Wilkins to take no more X-ray pictures of DNA.

If Watson and Crick wanted fresh pictures, they would have to get them from Franklin. Wilkins told the duo that Franklin was scheduled to give a seminar on her work in a few weeks, and invited Watson to attend. Having been in Cambridge for only a few weeks, Watson needed to bone up on his crystal-lography so that Franklin's talk would not fly over his head.

The key clue Watson was looking for was whether DNA was in fact a helix. Franklin's X-ray photographs were sharper than Wilkins', but she did not offer a firm interpretation on that score. Later, as Franklin wrestled with interpre-tations of the so-called crystalline or A form of DNA, she had doubts. She did discuss how she had been working on two forms of DNA, what she called the "crystalline" form that was prepared similarly to Wilkins' samples, and a "wet" form that was prepared and analyzed under higher humidity. Franklin had performed careful measurements on the water content of each DNA form, but that was about all Watson took away. Franklin clearly believed that more X-ray crystallography was the key to the mystery, and did not mention any model building of the type Watson and Crick were planning.

The next day Crick grilled Watson for the details of the talk. As was his habit, Watson had not made any written notes. Crick was annoyed that he had not recorded the exact water content of the DNA, which would affect the placement of water molecules in any model. Nonetheless, Crick started scrib-bling on some scratch paper and said that only a few structures could be compatible with the water values Watson recalled. Crick was confident that they could come up with a model—perhaps as soon as in a week or so.

Day by day the two men worked through the issues. The first was where to place the sugar–phosphate chains; they decided that they should go in the inte-rior of the molecule. That way, the different-sized bases would be facing out and would be easier to fit in the structure than if they were in the interior. Then they struggled with what held the chains together. Eventually, they came up with magnesium ions that could bridge phosphate groups between chains (although there were no data on what ions were present in the samples). But how many chains? They determined that three chains braided together best fit Wilkins and Franklin's X-ray measurements.

In just a matter of days, they had a model. Word spread through the build-ing. The key test would be to see how it fit Franklin's data. Crick invited Wilkins up to Cambridge for a look. Wilkins agreed to come the next day, and told Crick that he would be bringing Franklin.

Crick opened the discussion, but as he spoke, Franklin soon grew im-patient. She was not impressed by Crick's explanation and firmly disagreed that there was sufficient evidence that DNA was helical. Then, she quickly dismissed the magnesium ions in the model by pointing out that they would be surrounded by water molecules and could not form a tight structure. Worse, Watson had not remembered the correct water content. He was off by a factor of 24. Both he and Crick knew that if there were more water in the molecule, as Franklin said, the number of possible structures mushroomed.

Their breakthrough was a complete bust.

The consequences were more than just embarrassment. The Cambridge upstarts had trespassed onto King's College London's turf and flopped. The possibility of escalating tensions between the two government-supported research units did not seem worth the risk to Crick and Watson's superiors. The word came down from above that they were to abandon their work on DNA and leave it to the crystallographers in London.

Their quest was over almost as soon as it had started.

ON THE SIDELINES

There was no use in trying to appeal the ban. Crick's large personality grated on some in the research unit, and he had still not finished his PhD. It was wiser for him to keep his head down and get back to working on his thesis on the structure of hemoglobin. Watson, who had not even been two months in Cambridge before the triple-helix fiasco, needed to gain a more solid foundation in how to solve structures. He began work on the structure of a plant virus. For most of the next year (1952), the two largely stuck to other structures.

Franklin continued to work hard on the alternate forms of DNA and on getting better pictures. However, she found the tension at King's increasingly stressful and started looking for a new position elsewhere. She and Wilkins continued to avoid one another.

The London pair did not have DNA to themselves. Sooner or later, most believed, Pauling would turn his attention from his triumphal work on proteins to DNA. Watson and Crick learned that the great chemist was scheduled to come to a meeting in London in May 1952, and perhaps to visit the King's College crystallographers. The duo was eager to find out whether Pauling had in fact begun to tackle DNA. But at the last minute, Pauling did not show.

As his British hosts would learn, Pauling had his own problems—with the U.S. Department of State. Alarmed by the rapid spread of nuclear weapons, Pauling had used his eminence to become a vocal critic of the arms race between the United States and the Soviet Union. He urged for negotiations to curtail the power, spread, and testing of nuclear weapons. The U.S. government had become suspicious that Pauling was a communist sympathizer (he was not), and denied him a passport to attend the London meeting.

The action led to an international furor. Pauling had served the war effort admirably, even earning a commendation from President Truman. To deny him the opportunity to share his world-leading work with his peers was scandalous. It also meant that Pauling did not get a chance to visit with and hear the latest from Franklin, who had obtained some much better X-ray pictures.

Crick and Watson soon learned that another player in the game would be coming through Cambridge. Erwin Chargaff had published a series of papers on the composition of DNA. By working out methods for chemically separating and measuring the bases, Chargaff discovered that the ratio of the amounts of the four bases in DNA was not 1:1:1:1, as would be expected if DNA was a repeating sequence of four bases (ACGTACGT . . .) as had once been proposed. He had also determined that the base ratios were different in different species' DNA. Moreover, he had discovered that the ratio of the amount of adenine to thymine and the ratio of the amount of cytosine to guanine were about 1:1 each. Watson had read Chargaff's papers, Crick had not, by the time they sat down to lunch in one of Cambridge's dining halls.

The encounter did not go well. Chargaff was not impressed by what the pair had to say; rather, what did impress him was "their extreme ignorance." Chargaff realized that Crick was ignorant of what his chemical analyses had revealed.

"Well, of course, there is the 1:1 ratios," Chargaff said.

"What is that?" Crick replied.

"Well it is all published," Chargaff said, and went on to explain the ratios. In the course of the discussion, it became clear to him that Crick did not even know the chemical structures of the four different bases.

Chargaff later said, "I never met two men who knew so little—and aspired so much." The encounter reinforced Chargaff's opinion of "a typical British intellectual atmosphere, little work and lots of talk."

Chargaff offered no interpretation of the 1:1 ratios. Crick, however, was electrified. It immediately occurred to him that if there were some sort of complementary pairing of the bases in the structure, there would have to be 1 to 1 ratios among the bases. Soon afterward, Crick tried to figure out how the bases might pair by stacking the structures on top of one another. He could not figure it out and retreated back into his protein work.

While the pair were not working officially on DNA, the topic was never far from Crick and Watson's minds. In the summer, Crick ran into Rosalind Franklin at a meeting in Cambridge. In the line for tea, she told Crick that she was convinced that the form of DNA that Wilkins had photographed (the A form as it would be known) was not a helix. She thought it was perhaps an unwound version of the wet or B form, which she did believe was a helix, but had not spent much time analyzing. Crick was not convinced, and suggested to her that a pattern she was examining could be an experimental anomaly and should be dismissed. Franklin, however, continued to try to explain the result in terms of a rod, sheet, or even a figure eight of DNA.

Pauling finally received a passport and made it to Europe that summer, which gave Watson a chance to find out if he was, in fact, taking on DNA. Pauling gave no signs of interest in DNA, but Watson learned from Pauling's wife that their son Peter was coming to join their very own research unit in Cambridge. Peter was assigned a desk right between Watson and Crick. Both men enjoyed Peter's company, and hoped to be able to keep tabs on any developments out of Pauling's lab in California.

Just before Christmas, their hopes and fears were both realized. Peter received a letter from home in which his father mentioned that he had a structure for DNA. No further details were given, but Watson and Crick's hearts both sank. They shared the news with their bosses. Pauling had won the protein structure race, and now it seemed that the brilliant American had repeated the feat with DNA.

BACK IN THE GAME

The wait was agony. For several weeks there was no more news from California until one day at the end of January 1953, Peter walked in with a copy of his father's submitted manuscript. He announced that the structure was a triple helix with the sugar–phosphate backbone in the center. Watson was surprised, as it sounded very much like the structure they had built and abandoned a year earlier. Maybe they had it right after all? He snatched the manuscript from Peter's pocket and began reading.

Watson sensed right away something was not right. He studied the illustrations and realized that the phosphate groups were not ionized; they were bound to hydrogen atoms. This went against what Watson thought he knew about nucleic acids—in fact, with no net charge deoxyribonucleic acid would not be an acid at all. Pauling, the undisputed greatest chemist in the world, had blundered. Watson thought, "We are still in the game."

Crick shared the news with his bosses in the research unit, telling them that Pauling had it wrong, but would surely keep at it until it was right. Two days later, Watson went to London to tell Wilkins about Pauling's mistake.

Wilkins was occupied, so Watson went down the hallway to Franklin's lab. He tried to show her the manuscript and where Pauling had gone wrong. She objected right away to the mention of a helix, asserting that there was no evidence that DNA was a helix. Watson countered that a helix was the simplest structure, and that her dismissal of a helix was a mistake. Franklin was visibly aggravated, so Watson started to leave just as Wilkins appeared. Franklin turned away and closed the lab door behind her.

Wilkins confided to Watson that Franklin would soon leave King's College, and that she had begun to share her data in preparation for turning the work back over to Wilkins. Just days earlier, Franklin's student (and Wilkins' former student) Raymond Gosling had brought him one rather sharp X-ray image of an alternate form of DNA, the wet, B form. Gosling had obtained the image in May 1952 by wrapping the DNA fibers around a paper clip to keep them taut in front of the X-ray source. Watson asked what the images looked like, and Wilkins brought out a picture, numbered B51.

Watson's mouth fell open. The picture was simpler than images of the A form, with a sharp cross pattern in the center, a telltale sign of a helix (Figure 5.3). Over dinner, Watson tried to urge Wilkins back into the game by saying that Pauling would be back on the trail very soon. Wilkins was not persuaded—he wanted to wait until Franklin left. On the train ride home, Watson tried to sketch as much of the X-ray image as he could remember. He was determined to start making models again.

The next day, Watson ran into the heads of the Cambridge unit and told them about the picture and Wilkins' reluctance to get on with solving the structure. The answer seemed closer than ever, so he brought up the prospect of the British losing to Pauling again. This time, he did not get a restraining order—he was encouraged to get on with building models again.

THE DOUBLE HELIX

Watson told Crick what he had gathered from the photograph and from Wilkins about the B form of DNA—and that was a lot. The dark patches at the top and bottom of the photograph were produced by reflection of the X-rays off the bases. By knowing the distance and orientation of the DNA sample from the X-ray source and the film, the distance between the bases could be calculated to be 3.4 angstroms (1 Å = 1 ten-millionth of a millimeter). The spots that made up the cross pattern revealed that ten bases were stacked on top of one another in each turn of the helix. And from the

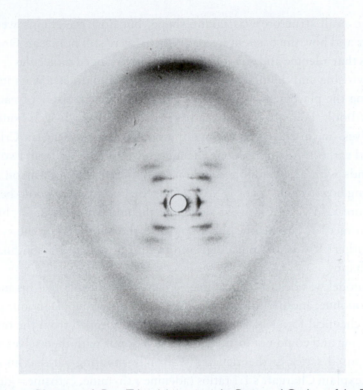

FIGURE 5.3 Photograph B51 Taken May 2, 1952, by Raymond Gosling of the B form of DNA, the X-ray diffraction pattern contains several key pieces of information. The closer the spots are, the greater is the actual distance between features. The pronounced crossing pattern of the spots is a diagnostic signature of a helical structure. The vertical distance from the center and the broad dark smears at the top and bottom corresponds to the 3.4-angstrom distance between stacked bases. The distance between horizontal rows of spots corresponds to the 34-angstrom distance between each turn of the helix. The two values together indicate that one turn consists of ten base pairs (34/3.4). King's College London Archives/Science Source.

diamond-shaped blank areas outside of the crossing pattern, it could be deduced that the phosphate backbone was on the outside of the helix.

The number of chains was not certain. That clue was found in a report in which Franklin had summarized her previous year's work for the British funding agency that supported it. The report was circulated among the research units, so Perutz (the head of the unit) received a copy. Crick and Watson asked to see it and Perutz shared it, since the report was not confidential. Crick quickly spotted a key data point whose significance had eluded Franklin: the structure had a 2-fold axis of symmetry, which meant that two chains ran in antiparallel, not parallel directions.

The overall structure was taking shape, but how the chains were held together and how the bases fit inside the helix were still not clear. One problem was that adenine and guanine were bulky, two-ringed molecules (purines) while cytosine and thymine were single rings (pyrimidines). Watson began exploring the placement of the bases. Studying the chemistry of each base in a textbook, he realized that the bases could form hydrogen bonds with themselves (in an end-to-end configuration, not by stacking as Crick had explored earlier). Adenine on one chain, for example, could form two hydrogen bonds with an adenine residue on the adjacent chain. Watson started imagining how two chains of DNA held together in this way would be identical (Figure 5.4A). As he worked alone into the middle of the night, his pulse began to race . . . Maybe one chain served as the template for the synthesis of the other? It was an appealing, even beautiful model. Watson finally fell asleep, happy.

His model was dead by lunchtime. When Watson explained his model to Jerry Donohue, another American in the Cambridge group, Donohue pointed out the chemical forms of the bases Watson had gleaned from the textbook were wrong and could not form the hydrogen bonds with themselves as Watson proposed. Crick further deflated Watson by pointing out that his scheme did not explain Chargaff's 1:1 ratios (adenine = thymine, cytosine = guanine). Watson had to go back to the drawing board—actually, to cardboard. Watson spent the afternoon making cardboard cutout models of each of the four bases.

The next morning, a Saturday, Watson was the first to arrive at the office. He cleared away a flat surface and began aligning the bases in various combinations. Suddenly he realized that an adenine–thymine pair held together by two hydrogen bonds had the same shape as a cytosine–guanine pair held together by at least two hydrogen bonds. He called over to Jerry Donohue to check his chemistry, and was ecstatic when Donohue had no objections.

Watson realized he had solved three problems at once. The base pairs would all fit nicely on top of one another without distorting the interior of the helix, the hydrogen bonds would hold the helix together, and the base pairs explained Chargaff's ratio. Moreover, the pairing rules suggested a way for copying DNA in that each chain was the complement of the other (Figure 5.4B).

When Crick arrived, he barely made it through the door when Watson told him they had the answer. The duo celebrated with lunch at the Eagle Pub, where Crick told everyone they had found "the secret of life."

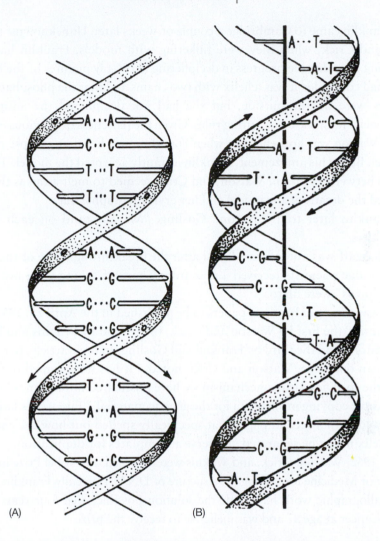

FIGURE 5.4 Two Models for Base-Pairing (A) Watson's first double helix model had a like-with-like base-pairing scheme. (B) Watson's second base-pairing scheme had adenine–thymine and cytosine–guanine base pairs that were the same size and fit Chargaff's ratios.

SHARING THE NEWS

Over the next two weeks, Watson and Crick assembled the physical double helix model of DNA. They invited Wilkins and Franklin to see it. Wilkins came first and liked it right away. He did not express a trace of resentment or regret that the Cambridge duo had solved the structure first.

Franklin came to Cambridge a couple of weeks later. Unbeknownst to Watson and Crick, while they were tinkering with models, Franklin had been making considerable progress in deciphering her X-ray pictures of the B form. She had concluded it was a helix with two chains and that the phosphate backbones were on the outside, but she had not deciphered the antiparallel arrangement or the base-pairing rules. Given the tenseness of previous encounters, Watson was apprehensive when Franklin took her first glimpse of their model. But to his amazement, Franklin instantly accepted the model. The tensions between Franklin, Watson, and Crick seemed to melt away as they discussed the details of the model and her crystallographic data.

Franklin later told Raymond Gosling, "We all stand on each other's shoulders."

Chargaff was not informed. Just after the breakthrough, one of the senior Cambridge scientists received a note from Chargaff asking what his "scientific clowns" were up to.

It was arranged for three papers to be published in the April 25, 1953, issue of *Nature,* the first by Watson and Crick, the second by Wilkins and his collaborators, and the third by Franklin and Gosling. In their article, barely longer than one page, Watson and Crick noted, "It has not escaped our notice that the specific pairing mechanism we have postulated immediately suggests a possible copying mechanism for the genetic material." Five weeks later, they followed with a second paper that specifically spelled out how the chains of the helix might be separated and serve as templates for duplication.

In 1962, Watson, Crick, and Wilkins were awarded the Nobel Prize in Physiology or Medicine for solving the structure of DNA. Tragically, Franklin, whose crystallographic work had made the solution possible, died four years earlier from cancer at age 37 and was ineligible to receive the prize.

END-OF-CHAPTER QUESTIONS

1. Which important clue or clues to the structure of DNA did each of the six main scientists obtain? Which key clues did each overlook or lack, and why?
   a. Erwin Chargaff
   b. Linus Pauling

   c. Maurice Wilkins

   d. Rosalind Franklin

   e. James Watson

   f. Francis Crick

2. Long before the structure of DNA was solved, the physicist Erwin Schrödinger suggested that the three-dimensional arrangement of atoms in some polymer had to explain the two main properties of heredity. Those properties were (i) the stability of life, such that traits were passed faithfully from generation to generation; and (ii) the mutability of life, such that traits could change.

   How did Watson and Crick's structure explain these two properties?

3. What problems existed with Watson's like-with-like base-pairing scheme?

4. Much has been written and said about the issue of credit concerning the discovery of the structure of DNA, particularly with respect to Rosalind Franklin. For each of the following accepted facts of the story, discuss the effect of each action on the solution of the structure of DNA, and whether the actions taken were appropriate or fair.

   a. Max Perutz shared Franklin's nonconfidential report to the British funding body with Watson and Crick (without Franklin's knowledge).

   b. Maurice Wilkins showed photograph B51, taken by Gosling and Franklin, to James Watson (without Franklin's knowledge or asking her permission).

   c. Watson and Crick's acknowledgments in their first report stated: "We have also been stimulated by a knowledge of the general nature of the unpublished experimental results and ideas of Dr. M. H. F. Wilkins, Dr. R. E. Franklin, and their co-workers at King's College, London."

   d. The Nobel Committee awarded the 1962 Nobel Prize in Physiology or Medicine to Watson, Crick, and Wilkins *"for their discoveries concerning the molecular structure of nucleic acids and its significance for information transfer in living material."* Franklin, who died four years earlier, was ineligible under the rules of the prizes.

6

GENES AS MEDICINE

Alone we can do so little, together we can do so much.
—HELEN KELLER

IN THE MEDIEVAL LEGEND of King Arthur, Lancelot was the most handsome and gallant of the Knights of the Round Table who joined the king in an epic quest to find the Grail—a vessel with miraculous powers.

In the annals of genetics, another Lancelot played a heroic role in the epic quest for one of the long-sought grails in modern medicine—gene therapy, and its powers to cure inherited diseases. This Lancelot, however, while also handsome and descended from medieval ancestors, was a sheepdog.

Lancelot made history as just a four-month-old puppy, when he and two of his siblings (named King Arthur and Guinevere) were treated for an inherited form of blindness. Some dogs of this breed, known as the briard, carry a recessive mutation that when present in two copies causes early and severe vision loss along with a slow degeneration of the retina. The affected puppies bump into things so often that they retreat to the edges of rooms and avoid moving around. Lancelot could not even find his water bowl.

On July 25, 2000, Lancelot was prepared for surgery at his home at the University of Pennsylvania's veterinary facility in New Bolton, outside Philadelphia. He was administered anesthesia and monitored closely before pediatric eye surgeon Albert Maguire injected just a few droplets of a solution

(150–200 microliters, about the volume of a pea) into the space below the retina of his right eye, leaving the untreated left eye as an experimental control. The solution contained about 40 billion copies of a specially engineered virus into which a functional version of the mutated gene had been inserted. Guinevere and King Arthur received the same treatment a couple of weeks later.

The large team of researchers involved in the experiment hoped that the virus would infect and deliver its genetic payload into enough cells in the dogs' retinas, and that the gene would work well enough, that they might be able to measure some quantitative change in the dogs' eyes within a few months. And they hoped just as much that there weren't any negative effects. All across the country, gene therapy experiments in humans had recently been halted by regulatory authorities out of safety concerns. Indeed, the once promising future of gene therapy was in serious doubt.

Just weeks after treatment, one of the animal technicians caring for the dogs phoned Jean Bennett, professor of ophthalmology and leader of the research team at Penn. The technician said, "Jean, these animals are watching us as we come into the control room. They used to just look off into space, and now they are following us as we move back and forth!"

Bennett promptly drove out to the kennels to see the dogs for herself. The change was dramatic. They no longer had their far-off gaze—they were clearly watching, and running around. Tests soon revealed that they had regained the ability to avoid objects, even to catch things, and that the pupils of treated eyes now responded to light.

Bennett was ecstatic. The experiment had succeeded far beyond anyone's expectations. She thought, "Wow, we can make blind puppies see. Wouldn't it be amazing to make blind children see?"

Curing humans—that would be Bennett's next goal. Indeed, that had been the quest to which Bennett and Maguire had already devoted nearly 20 years. And brave Lancelot would play a starring role in their eventual victory (Figure 6.1).

PARTNERS

Bennett and Maguire had been partners, in one form or another, since their first week of medical school in 1982.

Bennett arrived already knowing a lot of biology. She had previously earned a PhD in zoology from the University of California at Berkeley. She also knew

FIGURE 6.1 Lancelot Born with a mutation that caused early blindness, the dog was the first to be successfully treated by gene therapy, and later went to Washington to promote research on eye diseases. Photo courtesy of Foundation Fighting Blindness.

what kind of doctor she wanted to be. After her PhD and while doing research on early development in mammals, she visited the laboratory of W. French Anderson at the National Institutes of Health (NIH). Anderson was keen to use the newfound power of cloning and expressing mammalian genes to try to correct human genetic diseases—via "gene therapy." Bennett was very excited by the idea, and decided that was what she wanted to do. The problem was, she had worked only on sea urchins and mice. She peppered Anderson with questions.

"How can I become a gene therapist?" she asked.

"Simple. Just go to Harvard Medical School, learn about the diseases and then go back to the lab," replied the Harvard-trained Anderson.

Right—just apply to Harvard Medical School. What could be easier?

So Bennett applied . . . and she got in!

In neuroanatomy class, every pair of students was given a donated brain to dissect. She and Maguire happened to be assigned one another as lab partners. Bennett did not know much about brains, but Maguire had studied neuroscience as an undergraduate. He also knew what kind of doctor he wanted to be—he hoped to become a brain surgeon.

The pair got along well, so well that the lab partners were married by their third year of medical school.

Throughout their training, they frequently discussed the promise and challenges of gene therapy. Bennett introduced Maguire to the basic idea. She explained that to introduce a functional gene into a human was an extension of what was going on in labs around the world. The feat had been accomplished for fruit flies and mice in 1981–1982, just before they entered Harvard. It was a matter of translating that knowledge to humans.

Maguire thought the human step was a huge leap, and he was right. Mutations in hundreds of different genes caused various diseases in humans, but other than the mutated globin gene that caused sickle-cell anemia, no other specific human gene mutation had been identified. Moreover, no technology existed for delivering genes to cells inside the human body, let alone for getting genes to a specific organ or tissue, and the safety or efficacy of any eventual technology was yet to be seen.

As part of his rotation experiences after his second year of medical school, Maguire spent time with a specialist on retinal degeneration. The experience prompted Maguire to ask Bennett, "Do you think we could use gene therapy to treat retinitis pigmentosa?" (an eye disease that leads to blindness).

"Sure!" Bennett replied. Always the optimist, she did not tell Maguire at the time that nothing was known about the genes involved in retinitis pigmentosa or any other eye disease.

But the notion of tackling eye diseases was a turning point for the two medical students. The eye was an attractive target for gene therapy because it was more accessible than internal organs. Moreover, it was known to be an immunologically privileged organ, so foreign substances might be introduced into the eye without causing a damaging immune response. And eyes had the added appeal that everyone had two, so one eye could serve as a perfect control in any experiment.

Maguire decided to forego neurosurgery and to pursue training as a retina surgeon. Bennett decided that she would put off further clinical training, and focus on how to get genes into and expressed in the eye. After the birth of their first child—or as they told friends, their "senior project"—the couple received their medical degrees and left Boston for further training.

A THERAPY IN SEARCH OF A DISEASE

The basic ingredients necessary for gene therapy were clear to Bennett and everyone else contemplating the treatment of some disease. One needed: (i) a functional version of the gene that was mutated in a given disease; (ii) a way of getting that gene into the appropriate organ or cell type (e.g., lung, retina); and (iii) a means of making the gene active in those cells. When Bennett left Harvard in 1986, however, almost none of these essential ingredients were in hand for any human disease. Indeed, it was unclear which human disease might offer the best prospect for being treated with gene therapy. Bennett figured that before the details could be worked out for humans, it would be necessary, and prudent, to work out the science first in laboratory mammals.

She concentrated on finding a good animal model. Her earlier experience in developmental biology had made her an expert in making genetically engineered mice. While Maguire did his ophthalmology and surgery residencies, first at Yale, then at the University of Michigan, and finally at Johns Hopkins, Bennett collaborated with several researchers to identify the parts of genes necessary for their activation in mouse eyes.

In the meantime, there were rapid developments in human genetics. A milestone was reached in 1989 when the gene involved in cystic fibrosis (CF), the most common inherited disease in Caucasians, was identified (Figure 6.2). With the gene in hand, and because of the prevalence of the usually fatal disease, CF was a prime candidate for gene therapy.

But delivering genes into an organ within the body was particularly challenging. How could one specifically target the lungs, for example, of CF patients? Moreover, lungs are large organs, and how could one get a functional gene into enough cells to mitigate the disease symptoms?

One possible strategy for getting around the inaccessibility of internal organs was to consider diseases of blood cells, which could be removed from the body, treated in culture in the laboratory, and then returned to the patient. In 1990, W. French Anderson and colleagues at the National Institutes of Health took this strategy to attempt the first gene therapy in humans. The disease they tackled was a very rare form of severe combined immune deficiency (SCID) in which patients have greatly reduced numbers of T and B lymphocytes, and are thus vulnerable to all kinds of infections. The disease is caused by a defect in the adenosine deaminase (ADA) gene that encodes an enzyme responsible for removing deoxyadenosine from cells. The metabolite is particularly toxic to immature lymphocytes.

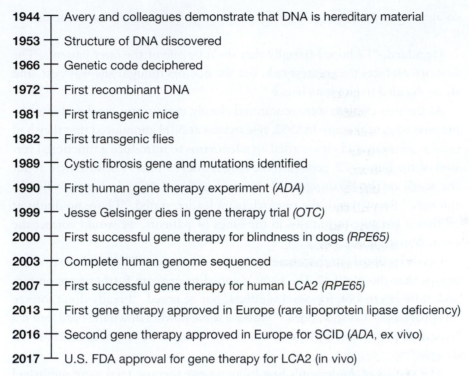

1944	Avery and colleagues demonstrate that DNA is hereditary material
1953	Structure of DNA discovered
1966	Genetic code deciphered
1972	First recombinant DNA
1981	First transgenic mice
1982	First transgenic flies
1989	Cystic fibrosis gene and mutations identified
1990	First human gene therapy experiment *(ADA)*
1999	Jesse Gelsinger dies in gene therapy trial *(OTC)*
2000	First successful gene therapy for blindness in dogs *(RPE65)*
2003	Complete human genome sequenced
2007	First successful gene therapy for human LCA2 *(RPE65)*
2013	First gene therapy approved in Europe (rare lipoprotein lipase deficiency)
2016	Second gene therapy approved in Europe for SCID *(ADA*, ex vivo)
2017	U.S. FDA approval for gene therapy for LCA2 (in vivo)

FIGURE 6.2 A Time Line of Milestones in Genetics and Gene Therapy

Existing treatment for the disease involved repeated injections of a long-lasting form of the enzyme obtained from cows. Anderson and colleagues hoped that delivering the ADA gene to lymphocytes would restore immune system function, and perhaps enable patients to no longer require enzyme treatments. To deliver the gene, researchers inserted a compact form of the human gene (without any introns) into a retrovirus that was capable of inserting into the cells' DNA. Two children, ages 4 and 9, had their T lymphocytes removed from their blood, exposed to the genetically modified retrovirus, and then returned to their bodies.

Just the start of the experiment, before any effects were known, garnered a great deal of press attention. Anderson told the *New York Times*, "We feel that gene therapy is potentially a major new medical option." He added, "And the most important thing, with any new therapy, is to get started. Now we're getting started."

The father of the 4-year-old, the first girl treated, said, "We'd been really excited and hopeful about gene therapy since we first heard about it a year or so ago. But we thought it would take 20 years before anything came out of it."

He added, "I'd hoped initially that she'd be one of the later patients. The first person faces the greatest risk. But the doctors thought she was best, and so we decided to give it a chance."

As the two children were monitored closely, enthusiasm escalated over the promise of gene therapy. In 1992, researchers at NIH announced that they had used a common cold virus called an adenovirus to introduce a functional version of the human CF gene into the lung tissues of rats. A news article about the work in the *Washington Post* was headlined "Cystic Fibrosis Breakthrough." Ron Crystal, the research team leader, stated, "I have no doubt at all that if we were to put this in the lungs of patients, we would correct the cystic fibrosis now."

Crystal pointed out that there was still work to do. "What we have to show now is that this is safe." The team planned to perform further tests on rats and monkeys to look for adverse effects, but he noted, "I really don't foresee any problems." Crystal added, "We have some hurdles down the road, but it's remarkable how fast this field is moving. I think we're going to cure this disease."

The results of Anderson's first human gene therapy trial were published in 1995. One patient achieved and maintained a normal white blood cell count, but the second patient did not. The engineered virus was detectable in nearly 100% of the first patient's T cells, but only about 1% of the second patient's T cells had incorporated the virus carrying the ADA gene. The reason for the different outcomes was unclear. Interpretation of the experiment was further complicated by the fact that the children's injections with the ADA enzyme were continued throughout the experiment. It was not possible to disentangle the specific benefit of the gene therapy from that of the standard treatment.

Nonetheless, the field brimmed with optimism. By the mid-1990s, scores of gene therapy trials were underway across the world. Medicine appeared to be on the brink of a revolution.

MICE ARE NOT HUMANS

Bennett and Maguire were not yet ready to try anything in humans. After moving to Penn in 1992, they first sought to establish the principle of gene therapy for an eye disease in animals. By 1996, they demonstrated that retinal degeneration in mice caused by a specific genetic mutation could be delayed by injecting an adenovirus that had been engineered to carry a functional version of the gene.

But they understood very well that mice are not humans. Mouse eyes are tiny in comparison, and the animals are not very visual creatures; they rely more on their sense of smell to navigate the world. Bennett and Maguire looked for better anatomical and behavioral models of human vision and disease, and decided that dogs would be a good next step. Maguire worked on developing and perfecting surgical techniques, while Bennett and collaborators investigated potential disease models, and the engineering of different kinds of viruses for carrying genes.

A syndrome in Irish setters appeared to be a promising target. The breed carried a mutation that caused a condition known as rod–cone dystrophy that was similar to the retinal degeneration Bennett and Maguire had treated in mice. Further study revealed, however, that the retina started deteriorating at such an early age that intervention did not seem feasible. It was back to the drawing board, or rather the kennel, to find a potentially treatable condition.

Fortunately, Bennett's veterinarian colleagues Greg Acland and Gus Aguirre managed colonies of dogs with retinal diseases. They and Kristina Narfström at the Swedish University of Agricultural Sciences in Uppsala had recently discovered the specific mutation responsible for blindness in briards. Importantly, both the gene involved, dubbed *RPE65*, and the retinal pathology turned out to correspond to a human condition known as Leber's congenital amaurosis 2 (LCA2). The *RPE65* gene encodes a protein that is necessary for the production of *cis*-retinal, a form of vitamin A that is bound by the opsin proteins that detect light in the eye's photoreceptor cells. When the RPE65 protein is missing, the response to light is impaired, and the retina degenerates over time. With the *RPE65* gene identified, viral delivery systems validated, and surgical techniques worked out for dogs, the disease appeared to be a very good model for gene therapy.

But before the experiments could get started, disaster struck.

HUMANS ARE NOT MICE

By the late 1990s hundreds of gene therapy clinical trials were underway, and Penn was a major hub of activity. Almost all of the studies were Phase I trials which, in accordance with the U.S. Food and Drug Administration (FDA) drug approval process, are small-scale studies to test the safety of a particular experimental treatment. Once safety had been demonstrated, efficacy would then be tested in larger numbers of people by comparing treated and untreated control patients in Phase II and Phase III trials.

One Phase I trial at Penn involved a potential treatment for a very serious, but rare, condition caused by mutations in the gene encoding the enzyme ornithine transcarbamylase (OTC). People with defects in the gene develop very high levels of ammonia in their bloodstream that can cause brain damage, coma, and death. The enzyme is normally active in the liver, so researchers sought to deliver a functional gene to patients' livers by infusing an engineered adenovirus carrying a functional version of the OTC gene directly into the main artery of the liver.

Jesse Gelsinger, an 18-year-old man from Tucson, volunteered for the trial after he heard about it from his doctor. He had suffered several bouts of high levels of ammonia, including a coma in late 1998. Gelsinger traveled to Philadelphia and on September 13, 1999, he became the eighteenth person in the trial, and the second to receive the highest dose of virus, over 38 trillion particles.

Just 12 hours later, Gelsinger was feeling unwell and spiked a fever of 104.5 degrees. The previous seventeen subjects had similar symptoms the first day. But by the next morning, Gelsinger was disoriented and his skin and the whites of his eyes had become yellow. The jaundice was a sign of liver trouble. Ammonia levels in his bloodstream then climbed to about 10 times their normal level. By the following morning, Gelsinger had fallen into a coma and needed assistance breathing. Doctors used dialysis to remove the ammonia, but Gelsinger continued to decline as blood clots formed in his bloodstream. Gelsinger's family raced to Philadelphia to be at his side. The doctors and staff were powerless, however, as his organs began to fail. Gelsinger died just 98 hours after the start of his treatment.

The tragedy sent shock waves throughout the Penn medical community and the field of gene therapy. Gelsinger's doctors, all colleagues of Bennett and Maguire, were devastated. How could they have lost this young brave patient? What went wrong?

It would take months to find out. In the meantime, as a precaution, the NIH put a hold on all gene therapy trials while investigators examined not only the Gelsinger case, but the conduct of other trials as well. Investigations by the NIH, the FDA, and Penn revealed a field in a hurry for results. Serious reactions to the infused virus had been observed in animal experiments, but had not raised great enough concerns to delay human testing. Worse, the potential risks had not been fully divulged to potential human volunteers, including Gelsinger. Investigators also discovered that a very large number of adverse reactions in other trials, including deaths, had not been reported to regulatory authorities.

Paul Gelsinger, Jesse's father, testified before a Senate subcommittee, tearfully describing his son's death as "an avoidable tragedy from which I will never recover." The consensus was that proponents had overpromised the potential benefits while underplaying the risks.

As government agencies worked on new guidelines, the FDA shut down all gene therapy trials at Penn, and the entire field came to a grinding halt. Some magazines ran stories with headlines such as "Gene Therapy R.I.P." No one could say when, or if, human trials would be restarted. It was clear that certain viral vectors had to be modified or replaced, and that more extensive animal testing would be required before human trials could resume.

Then, just when the future of gene therapy looked very bleak, a shaggy sheepdog stepped onto the national stage.

LANCELOT GOES TO WASHINGTON

The cautious, one-step-at-a-time process that authorities demanded in the wake of the Gelsinger tragedy was exactly the approach that Bennett, Maguire, and their team had embarked on years earlier. They had already moved to working with a different virus called adeno-associated virus (AAV) that, unlike adenovirus, was not pathogenic. They had also discovered that AAV was more efficient at targeting retinal cells than adenovirus. And very importantly, they had demonstrated that gene transfer with AAV lasted for years, and that the virus could be readministered without provoking a strong immune response.

There was no moratorium on animal experiments. So, in the summer following Gelsinger's death, Bennett, Maguire, and their collaborators proceeded with the first experiments on Lancelot, King Arthur, and Guinevere. Their dramatic success in making the blind dogs see drew enormous press attention, and made Lancelot a star attraction.

Lancelot made his national television debut on *Good Morning America*. And to help boost the case for funding gene therapy in the aftermath of the Gelsinger investigations, he made three trips to Congress. Bennett accompanied Lancelot on his first trip in May 2001, where she spoke to a packed room at a formal luncheon. Bennett was worried that Lancelot would be timid or scared, but he sat perfectly still as she gave her talk and showed movies of Lancelot before and after gene therapy. One congressman dropped his napkin, and people noticed Lancelot followed it with his treated eye. Others then dropped their napkins so Lancelot could demonstrate his ability to see. The happy pooch so charmed lawmakers that he made it into the official *Congressional Record*, which noted: "This recent National Eye Institute–supported research that has given sight to Lancelot holds promise for children born blind."

Bennett was deluged with calls from parents and patients eager to try the therapy. But there were issues that needed to be addressed in dogs before human testing could be considered. Foremost among them were the duration of improved vision and the reproducibility of the treatment. The team treated another 14 dogs (26 eyes in all) and found that just a single treatment could substantially restore vision that remained stable for up to three years—a long time in dog years. The results bolstered hope for the treatment of humans, whenever that might be.

CHANGING GENES, CHANGING LIVES

On July 25, 2005—five years to the day after Lancelot's treatment—Bennett's colleague Kathy High walked into her office, sat down, and asked, "Jean, how would you like to run a clinical trial?" High had been working diligently since the Gelsinger episode to assemble a team of experts at Penn that could produce clinical-grade AAV virus, design clinical protocols, and prepare the documents required by regulatory authorities for gene therapy trials. The next step was to conduct an actual trial and, based on the dog data, the treatment of LCA2 was the most promising candidate.

Bennett was thrilled. She and Maguire hurled themselves into the task. This was to be the first trial to include children with LCA2, to be conducted in parallel with other Phase I trials on adults with LCA2 in Philadelphia, Florida, and London. In October 2007, Maguire treated the first of three adult patients. Within a few weeks of being administered the genetically engineered virus, two patients who were previously able only to see hand motions were able to read several lines of an eye chart. The Florida and London trials also reported early success.

Maguire and Bennett expected, however, that the improvement in children might be even better since the disease would not have progressed as far in their younger eyes. In September 2008, Dr. Al, as his pediatric patients called him, treated the first child, an 8-year-old boy, the youngest that regulations would permit. Just four days later, on a visit to the Philadelphia Zoo, the boy startled

FIGURE 6.3 Jean Bennett and Albert Maguire with Pet Briards Venus and Mercury Both dogs were treated for early blindness by gene therapy, and then adopted by the researchers. Brent Stirton/Getty Images Reportage.

his parents when he tried to gaze up at the sky—and complained that the sky was *too bright*.

The boy's parents phoned Jean Bennett in tears. *It worked.* Their son would soon be able to see his friends' faces for the first time, to see lightning bugs, and to read his schoolbooks.

For Bennett and Maguire, the experience was no less emotional, as Bennett later recalled:

> It was like witnessing a miracle. To be able to see a child . . . who came in with a blind cane because they couldn't see, and then to see that child running around and riding a bicycle and reading books—doing things that normal children do. It transformed their lives. I feel like one of the luckiest people in the world to be able to witness that. And those people are the real pioneers.

All twelve patients in the Phase I trial showed improvement. For the Phase II part of the trial, the same twelve patients had their other eye treated. Twenty-one additional patients were treated in the Phase III trial.

On December 19, 2017, thirty-five years after Bennett and Maguire first teamed up in medical school, the U.S. FDA formally approved gene therapy for LCA2, the first in vivo gene therapy approved in the United States (Figure 6.3).

END-OF-CHAPTER QUESTIONS

1. (a) What information and ingredients are needed to attempt gene therapy of an inherited disease? (b) Were all of these ingredients in hand for the treatment of OTC? What went wrong in the case of Jesse Gelsinger?
2. Are animal models necessary for human biomedical research? Justify your statement.
3. What did researchers learn about gene therapy of the *RPE65* mutation in briard dogs that was important for treatment of humans with LCA2?
4. The gene that is mutated in cystic fibrosis patients was identified in 1989. As of today, there is no gene therapy for this disease. Why were researchers so confident about curing the disease in the 1990s?
5. Gene transfer was established in animals in the early 1980s, but the first approved human gene therapies did not occur for more than 30 years. Why did the application of basic knowledge about genes to human therapy take so long?

Part III

EVOLUTION AND THE ORIGINS OF BIOLOGICAL DIVERSITY

FOR THOUSANDS OF YEARS, humans of all cultures have asked, "Where did I come from?" and "How did all the creatures around me come into being?" And they created all sorts of supernatural explanations. It was not until 1859 (fairly recently in the larger scheme of human history), when Darwin published *On the Origin of Species*, that a convincing explanation for the natural origins of creatures, including humans, was put forth. The making of that revolutionary theory is also perhaps one of the greatest stories in all of science (Chapter 7), as it involved both great adventures and great thinking, and not just by Darwin. While Darwin's voyage and labors are well known, and rightly so, the making of the theory of evolution also owes a considerable debt to Alfred Russel Wallace, who undertook two long journeys after Darwin, witnessed similar patterns in nature, and arrived at very similar ideas.

One of the most important impacts of the theory was to prompt later scientists to explore some of the larger gaps in our knowledge about the origins of various groups. The succeeding five stories in this section are each about pioneers who made breakthroughs into some of the deepest and most compelling mysteries in the history of life, including the origins of animals (Chapter 8), the origins of eukaryotes (Chapters 9 and 10), and the origins of humans (Chapter 11), along with the very recent discovery of our Neanderthal ancestry (Chapter 12).

7

GREAT MINDS THINK ALIKE

*On a careful consideration, we find a curious series of corre-
spondences, both in mind and in environment, which led
Darwin and myself, alone among our contemporaries, to
reach identically the same theory.*

—ALFRED RUSSEL WALLACE (1908)

BY THE TIME HE reached São Joaquim, deep in the heart of the Amazon
and more than 1500 miles upriver from the Atlantic Ocean, Alfred Russel Wallace was near collapse. Crippled by a recurring high fever and severe chills, he
had no appetite and became so weak he could not even turn over in his
hammock, let alone lift himself. A friend who was looking after him did not
expect the 29-year-old British naturalist to survive.

But regular doses of quinine, the antimalarial extract from the bark of
the cinchona tree, and two months of rest made Wallace just strong enough
to be able to walk down to the river using a stick as a cane. Figuring that he
could continue to recuperate just as well in a canoe as in a hammock, he
decided to resume his Amazon journey, now in its fourth year. On February 16, 1852, accompanied by a team of local men, Wallace started his ascent
of the Rio dos Uaupés, a tributary that branched off from the Rio Negro
(Figure 7.1).

The river was full of rocks and rapids, with many waterfalls. The canoe
had to be either pulled through the rushing water with ropes from the shore
or unloaded and carried up along the margins of the falls. By March 12, they
had crossed over 50 rapids, 12 of which were so rough and had such high water
that the canoe had to be pulled by as many as 24 men. Still weak and

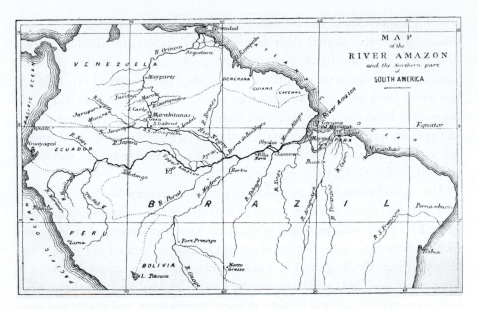

FIGURE 7.1 The Amazon Wallace traveled from Pará on the east coast of Brazil up the Rio Negro to the Rio dos Uaupés at the far western reaches of the river system.

From Wallace (1969).

suffering from bouts of fever, Wallace decided at Mucúra he had gone far enough. Neither his health nor his luck on the river were likely to hold up much longer. After two weeks' rest, he would turn around and descend the same treacherous waters, and then attempt the very long journey to the Atlantic coast in the hope of catching a ship back to England.

What would possibly drive a person to put himself through such misery and danger, in a strange, sometimes hostile country far away from home and loved ones?

There was, of course, the thrill of adventure, of going where few or no Europeans had gone before. The Amazon certainly offered that. And there were pleasures such as bathing in a clear river at midday "when dripping with perspiration . . . you can have no idea of the excessive luxury of it," Wallace wrote home early in his journey. But for Wallace, there were two more specific driving motives for exploring the Amazon.

The first was collecting. Wallace developed a keen interest in nature on solitary rambles through the British countryside. He was spurred into collecting when he met and befriended enthusiastic amateur entomologist Henry Walter Bates while teaching in Leicester. The two men spent their spare time building collections of the creatures around Leicester, particularly the local insects. They also shared their favorite books, especially travelogues that described the natural riches of faraway lands, such as Charles Darwin's account of his voyage around the world on the HMS *Beagle* (1831–1836). It was a new book about an expedition up the Amazon that inspired Wallace to propose to Bates that they undertake their own voyage to the Amazon to collect specimens of birds, mammals, and insects, both for themselves and to sell to other collectors. As neither man had any money, the sale of specimens was crucial for paying their expenses.

The second motive was to tackle a very great mystery. Wallace and Bates were well read on the scientific issues of their day, and the outstanding question in the mid-1840s was the origin of the great variety of species that collectors such as themselves were documenting in England or around the world. The long-held view by clergy and by most scientists of the day (most of whom in Britain were also ordained clergy of the Church of England) was that species had been specially created by God in their present form and place, and were unchangeable. A few naturalists (some writing anonymously) had dared, however, to suggest the idea of *transmutation*, that one species could change, or transmutate, into another. Because how, when, or if transmutation occurred was not at all clear, and many found the idea objectionable, proponents of transmutation were publicly criticized.

Yet, privately, the question of the origin of species vexed all sorts of scientists—geologists, biologists, even astronomers. The scientific revolution was in full bloom, and many laws of physics and chemistry were being discovered. The question of whether the making of new species was also a product of natural processes, or a divine miracle, was considered the supreme question of biology—"the mystery of mysteries" as one scientist put it.

Wallace suggested to Bates that their expedition could shed light on this very mystery:

> I begin to feel rather dissatisfied with a mere local collection (of insects)—little is to be learnt by it. I should like to take some one family, to study thoroughly—principally with a view to the theory of the origin of species. By that means I am strongly of the opinion that some definite results might be arrived at.

Wallace and Bates arrived at Pará on the coast of Brazil in May 1848. After a time exploring the region together, they split up, and Wallace headed up the main trunk of the Amazon, then up the Rio Negro and the Rio dos Uaupés tributary. In addition to thousands of preserved specimens, Wallace had accumulated a small menagerie of live animals—monkeys, macaws, parrots, and a toucan—that he hoped to take all the way to the London Zoo. Their upkeep was also draining him of what little energy he had remaining.

Three months after leaving Mucúra, Wallace made it back to Pará in July 1852. He found a ship headed for England, the brig *Helen*, and boarded it with about 20 animals, many boxes of specimens and notes, and set sail for home.

Four weeks into the journey and about 700 miles east of Bermuda, the captain came to Wallace's cabin and said, "I'm afraid the ship's on fire; come and see what you think of it." Wallace followed the captain to the hold and saw smoke pouring out of it.

The crew tried but could not douse the smoldering blaze. The captain ordered down the lifeboats. Wallace went to his hot, smoky cabin and salvaged a small tin box and threw in some drawings, some notes, and a diary. He grabbed a line to lower himself into a lifeboat, slipped, and seared his hands on the rope. His pain was compounded when his injured hands hit the salt water. Once in the lifeboat, he discovered it was leaking. Wallace watched his animals perish, and then saw the *Helen* burn to the waterline before sinking with a small fortune's worth of specimens.

Day after day passed in the open boats. Wallace was blistered with sunburn, parched with thirst, soaked by sea spray, exhausted by constantly bailing water, and near starvation. Bobbing in the middle of the Atlantic Ocean in a leaky lifeboat, unsure whether he would survive, at some moment Wallace must have pondered how he wound up in such a predicament. Since one key part of his quest was to search for the origin of species, there was someone whom he might have blamed. The irony of the situation was that, unknown to Wallace, the answer to the mystery of the origin of species was already known long before he ever set foot in the Amazon.

Charles Darwin knew that species changed. He knew from his voyage around South America *more than fifteen years earlier*—he had just not said so.

AN UNLIKELY REVOLUTIONARY

Darwin was just 22 years old when he boarded the HMS *Beagle* just after Christmas 1831. The invitation to join the voyage was a complete surprise. Darwin had received his undergraduate degree from the University of Cambridge the previous spring, a degree that required his pledge of adherence to the Thirty-nine Articles, the basic doctrine of the Church of England. He was planning to pursue a year of divinity studies with the idea of becoming a vicar. But Professor John Henslow recommended Darwin to the ship's Captain Fitz-Roy as a budding naturalist who could catalog plant, animal, and rock samples on what was supposed to be a two-year surveying voyage around the world.

An avid amateur collector with a keen interest in geology, Darwin leaped at the chance to see the Tropics he had read and dreamed about. Once his father was persuaded that the journey was not overly dangerous, or a waste of time, Darwin had to prepare quickly. His family was financially well-off, so Darwin was able to buy a variety of instruments as well as a new pistol and rifle. When he inspected the ship, the *Beagle*, it was a bit of a shock, just 90 feet long and 24 feet at its widest. It had just two tiny cabins. The six-foot-tall Darwin had to stoop to enter his quarters that he was to share with a 19-year-old officer and a 14-year-old midshipman. Darwin would sleep in a hammock slung over a chart table, just two feet below a skylight.

He brought aboard a few books, including a new treatise on geology, and a copy of the Bible, which he quoted as the authority on matters of morality. Darwin believed that species were fixed and unchangeable. He did not set out with any speculative idea or revolutionary theory in mind. Those thoughts were years and thousands of miles away. But the sights and specimens that would provoke and inspire Darwin started to appear early in the voyage.

In September 1832, while exploring the coast of Argentina south of Bahía Blanca, Darwin found some rocks containing shells and the bones of large animals. He used his pickax to free what he guessed were parts of some rhinoceros-like animal. The next day he found a large skull embedded in soft rock and spent so many hours removing it, he did not get back to the ship until after dark. Two weeks later, he found a jawbone and a tooth that he thought belonged to a *Megatherium*, or giant ground sloth. He was not sure of what all he had, but he crated the bones for shipment ("cargoes of apparent rubbish," Captain FitzRoy teased) so that the experts back in England could decipher their identities.

Eventually, it would be determined that Darwin had found remnants of seven species, including: a giant armadillo-like creature called a glyptodon; *Toxodon*, an extinct relative of the capybara; and three types of ground sloth—*Megatherium*, *Glossotherium*, and *Mylodon* (Figure 7.2). All of the fossils were from animals that were much larger than the most similar species living in South America. Their distinctly larger size would spur Darwin to ponder the relationship between extinct and living species.

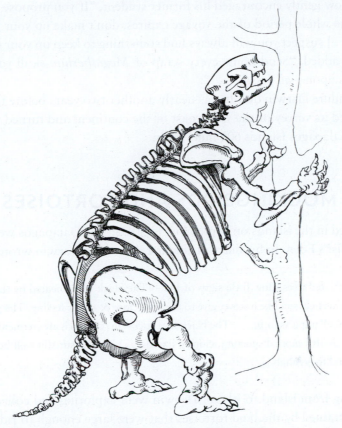

FIGURE 7.2 The Mylodon Darwin found the fossil remains of this giant ground sloth on the Argentina coast.

From Darwin (1890).

These treasures, however, came at a high personal price. Darwin never conquered his seasickness. As the *Beagle* made its way around the South American coast, Darwin was subjected to fierce storms and tossing seas. He took every opportunity to get off the ship and explore inland. Those opportunities were frequent, as the *Beagle*'s progress was very slow, much slower than expected. Two full years into the voyage, the ship was still surveying the east coast of the continent. Often knocked flat with seasickness and occasional bouts of homesickness for his family and the comforts of England, Darwin doubted whether he could withstand the length of the voyage. "I know not, how I shall be able to endure it," he wrote to Professor Henslow.

Henslow gently encouraged his former student, "If you propose returning before the whole period of the voyage expires, don't make up your mind in a hurry. . . . I suspect you will always find something to keep up your courage." Then he added, "Send home every scrap of *Megatherium* skull you can set your eyes upon—*all fossils*. . . ."

But endure Darwin did. It was nearly another two years before the *Beagle* completed its survey of the west coast of the continent and turned westward for the Galápagos Islands 600 miles away.

MOCKINGBIRDS AND TORTOISES

He arrived in the islands on September 15, 1835. The Galápagos were hardly a naturalist's Eden. In his diary of the first days there Darwin wrote:

> The stunted trees show little signs of life.—The black rocks heated by the rays of the vertical sun like a stove, give to the air a close & sultry feeling. The plants also smell unpleasantly. . . . The black lava rocks on the beach are frequented by large (2–3ft.) most disgusting, clumsy lizards. . . . They assuredly well become the land they inhabit.

Moving from island to island, Darwin went exploring and collecting. He was entertained by the land tortoises that were large enough to ride on. On James Island, Charles gathered all of the animals and plants he could. He was curious to decipher whether the plants were the same as those on the South American continent or peculiar to the islands. He also paid attention to the birds. The species of mockingbird on James Island looked different from those on other islands. So did the land tortoise. The primary goal, though, was collection—identification would come later.

After five weeks of hiking across hot black rock, Darwin was happy to leave. The *Beagle* sailed west and made stops in Tahiti, New Zealand, Australia, the Cocos Islands, and South Africa. All Captain FitzRoy had to do was to then turn north and sail home to England. But he didn't. Anxious to recheck some measurements, FitzRoy sailed across the Atlantic again to Brazil. Exasperated, Darwin wrote to his sister, "I loathe, I abhor the sea, & all ships which sail on it."

Reconciled to weeks more at sea, Darwin decided to make use of the time on board. His plan was to have experts back home study his collections of

FIGURE 7.3 Mockingbirds of the Galápagos Islands Darwin found three very similar but distinct species on four different islands. The slightly different species suggested to Darwin that one species had given rise to several forms, that species did in fact change.

From Darwin (1838–1841). Reproduced with permission from John van Wyhe, ed. 2002. The Complete Work of Charles Darwin Online (http://darwin-online.org.uk).

plants, animals, fossils, and rocks. He began to organize and write out his ornithological notes, which made him return to a puzzle posed by the birds and tortoises of the Galápagos. The mockingbirds were not especially attractive, but Darwin noticed that three different kinds occurred among four islands (Figure 7.3). He wrote in his notebook, "I have specimens from four of the longer islands; the specimens from Chatham & Albemarle Isd. appear to be the same, but the other two different. In each Isd. each kind is exclusively found; habits of all are indistinguishable."

The tortoises on different islands also looked different. Darwin continued his notes.

> When I see these Islands in sight of each other and possessed of but a scanty stock of animals, tenanted by these birds but slightly differing in structure & filling the same place in Nature, I must suspect they are only varieties. . . . If there is the slightest foundation for these remarks, the zoology of Archipelagoes—will be well worth examining; for such facts would undermine the stability of species.

Species might change—it was Darwin's first eureka moment.

The powerful idea gripped him. Finally returning to England after almost five years of voyaging, Darwin began to review all he had seen through the prism of this possibility. As experts pored over his collections, he opened the first of a series of private notebooks, Notebook B, on the transmutation of species. He jotted down his thoughts as they occurred to him.

He thought about the relationship between the extinct animals he found buried in the ground, and the living ones walking on top of them. On page 20, he wrote:

> We may look at Megatherium, armadillos, and sloths as all offsprings of some still older type. . . .

On page 21, he continued:

> Organized beings represent a tree *irregularly branched* some branches far more branched. . . .

On page 35, he noted:

> Similarity of animals in one country owing to springing from one branch. . . .

And on the next page, after writing "I think," he sketched a branching tree that depicted these ideas. It is the most famous drawing in natural history because it was a new system of natural history—one in which species are descended from ancestors, just as naturally as children are descended from parents and grandparents (Figure 7.4).

Life was a like a family tree. Darwin wondered, what made the tree branch? How did new species arise?

The questions, and possible answers, burned through Darwin's mind in a process he called "mental rioting." In September 1838, he opened Reverend Thomas Malthus' *An Essay on the Principle of Population* (1798). Malthus said that there were checks on the growth of human populations—disease, famine, and death—without which we would double in number every 25 years. Darwin recognized that the struggle to survive was even more intense in nature. Many species produced large numbers of offspring, of which few survived to adulthood and to reproduce. So what determined which offspring survived and which didn't? The answer came as a lightning bolt to Darwin—the stronger, better adapted ones survived, while the weaker perished. This winnowing process he would later call *natural selection*. The effect of natural selection, over time, would be the formation of new species.

Only two years after completing his voyage, Darwin had a new theory for the natural origin of species. And he had the perfect opportunity to reveal it. He wrote an account of his voyage based on his diaries, to be published along

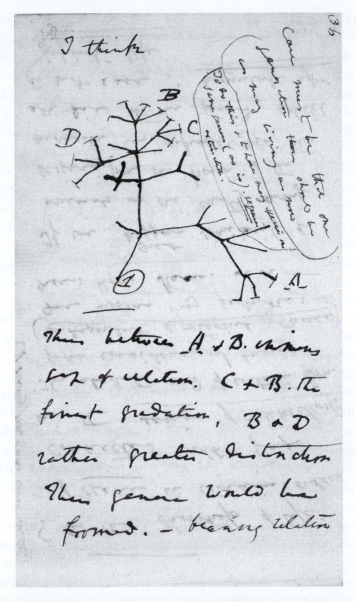

FIGURE 7.4 A New System of Natural History Darwin's entry on page 36 of his B notebook includes a drawing of a tree of life in which living species are descended from earlier species.

Reproduced by kind permission of the Syndics of Cambridge University Library, DAR.121.

with FitzRoy's and a former captain's reports on the *Beagle*'s surveying missions. Darwin described the places he visited, the colorful people he encountered, the phenomena he witnessed (including a terrifying earthquake), and the creatures he collected. When he came to the section of the book about the Galápagos Islands, he understood from experts he had consulted that the mockingbirds and finches on different islands were distinct species. As Darwin had surmised, they were similar to, but distinct from, mockingbirds and finches on the South American mainland. He deduced that some mainland birds had emigrated to the islands and over time produced new species. Darwin wrote, however, "It is clear that if several islands have each their peculiar species of the same genera, when they are placed together, they will have a wide range of character. But there is not space in this work, to enter on this curious subject."

He dodged the central issue completely. He avoided it because to divulge his conclusion that the origin of species was natural, and not divine, was heresy. It flew in the face of what he had been taught at Cambridge, what was held dear by his mentors who made the voyage possible and who were now introducing him to the British scientific elite. He would be spitting in their faces, and sink his personal, family, and professional reputations.

He had a bold new theory, but he did not dare tell anyone.

Darwin continued to accumulate all sorts of evidence that bolstered his conviction that he was on the right track. He saw the process of natural selection operating similarly to the process of domestication, with slight variations among individuals being favored by natural selection just as differences among domestic animals are selected by breeders. Darwin also thought that the phenomenon of vestigial structures, such as the wing bones of flightless birds, or the limb rudiments of snakes, revealed their descent from ancestors that possessed fully functional appendages. In 1842, Darwin distilled his notes and several years of thinking into a 35-page sketch to explain how "through the process of gradual selection of infinitesimal changes, endless forms most beautiful and most wonderful have been evolved." Two years later, he expanded this into an essay of 230 pages. But Darwin still thought it was unwise to publish. He divulged his ideas to only a few colleagues, and even then he was very cautious.

In 1844, he offered a morsel to botanist Joseph Hooker, who would become a trusted confidant:

> At last gleams of light have come, & I am almost convinced (quite contrary to opinion I started with) that species are not (it is like confessing a murder)

immutable. . . . I think I have found out (here's presumption!) the simple way
by which species become exquisitely adapted to various ends.

Rather than publicly confessing his murder of special creation, Darwin
remained silent on the subject of the origin of species. When Wallace read
what Darwin had written in his account of the voyage of the *Beagle*, he saw
the mystery as still unsolved. Had Darwin disclosed his theory in 1839, or
1842, or 1844, Wallace may not have gone to the Amazon in 1848. Wallace then
would not, of course, have wound up in that leaking lifeboat. But Darwin said
nothing, and so there Wallace was, stranded in the middle of the ocean. And
had Wallace not been rescued, we may have never heard of him. But he was
rescued, and his quest was far from over. And it would be Wallace who would
force Darwin to go public.

RESCUE AND RESOLVE

On the afternoon of their tenth day afloat, Wallace and the crew of the *Helen*
were spotted and picked up by a passing British ship bound for London. Wallace was saved, but once the danger had passed, the full weight of his loss
struck him. During his travels in the Amazon, his desire to carry on was
stoked by the prospect of bringing back new and beautiful species. Now, he
had nothing to show for his four years of hardship and sacrifice.

On the voyage home, Wallace composed a letter to a friend in Brazil detail-
ing his ordeal. He wrote, "Fifty times since I left Pará have I vowed, if I once
reached England, never to trust myself more on the ocean." Then he added,
"But good resolutions soon fade." Wallace decided that, despite his loss and
nearly deadly shipwreck, he would voyage again.

As he recovered his strength in England, Wallace pondered possible desti-
nations. He had to collect specimens that would fetch good prices. He ruled
out a return to the Amazon. He began thinking about the Malay Archipelago, the vast group of islands between Southeast Asia and Australia. Other
than on the island of Java, the animals and plants of the vast region—an area
almost as large as the entire continent of South America—were unknown to
European science. Covered in tropical forest, the islands often appeared similar, but they held different kinds of animals, and discovering and explaining
those differences would reward Wallace's perseverance and courage.

Wallace arrived on the island of Singapore in April 1854. Over the next eight
years he would make 96 crossings from island to island in a 14,000-mile

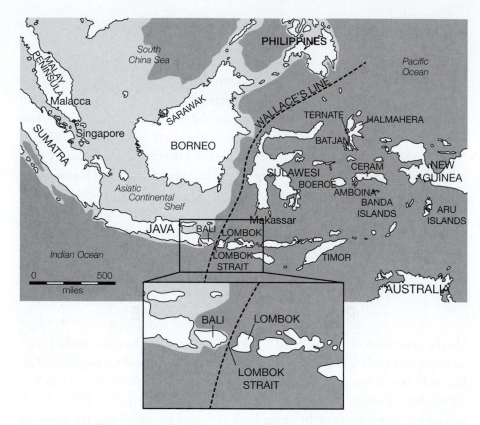

FIGURE 7.5 The Malay Archipelago Starting from Singapore in 1854, Wallace traveled for eight years across the islands collecting all sorts of animals. Map by Leanne Olds.

journey, and collect over 120,000 specimens that, this time, would make it home safely (Figure 7.5). Despite some reputation for ferocity, the local tribesmen shared their knowledge of the forest with Wallace and helped him to find what he was after. He stalked its most beautiful and prized specimens— orangutans, monkeys, spectacular birds of paradise, and enormous brilliant butterflies.

Coveted for their large wingspan and rich coloration, these so-called birdwing butterflies lured Wallace. Not only did he discover new species, but he also paid close attention to where he found the beautiful insects. He discovered that different birdwing types were restricted to particular islands. The related but distinct butterflies found on different islands signaled to Wallace just what the similar but distinct birds of the Galápagos Islands signaled to Darwin—that species change (Figure 7.6).

FIGURE 7.6 Birdwing Butterflies These large butterflies were prized by collectors. This sample is from Wallace's personal collection that he brought back from his travels.

The Natural History Museum/Alamy Stock Photo.

But while Darwin was keeping quiet about evolution, Wallace had no such reservations. He had a reputation to make and, lacking any family wealth or status, nothing to lose. He put his thoughts on paper and dashed them off to scientific magazines and journals in England. Some of these dispatches were short field notes; others revealed bigger ideas.

Having explored jungles on opposite sides of the globe, Wallace had the special opportunity to compare different groups of animals, and to ask why they were found where they were. Birdwing butterflies only occurred near other species of birdwings in the Malay Archipelago, whereas different families of butterflies lived in the Amazon. Similarly, he saw macaws only in South America, and cockatoos were found only in the Malay Archipelago and Australia. Around the globe, the more similar two species were, the closer they tended to live. Wallace cited the Galápagos animals Darwin described as demonstrating the same phenomenon. Wallace pointed out, however, that their distribution had not "received any, even a conjectural explanation" (because

Darwin had dodged the subject). Wallace asked, "Why are these things so? They could not be as they are, had no law regulated their creation and dispersion." These patterns of the distribution of animals suggested to Wallace that species must come from preexisting species. Writing from Sarawak on the island of Borneo, Wallace coined his Sarawak law: "*Every species has come into existence coincident both in space and time with a pre-existing closely allied species.*"

Wallace summoned more evidence from the presence of vestigial limbs in snakes, and finger bones in the flippers of manatees and whales. He wrote, "To every thoughtful naturalist the question must arise, What are these for?" These useless bones made no sense if each species were specially created without any regard to previous species, designed to fit its habitat and lifestyle. No, Wallace asserted, these vestigial structures teach us that species are connected through time and in space, like "a branching tree." On his own, Wallace had arrived at Darwin's concept of the branching tree of life.

Hopping across the Malay Archipelago, Wallace found more evidence for his law. In the jungle on the island of Borneo, Wallace saw monkeys and orangutans. But as he traveled farther east to the jungles of New Guinea, he did not find monkeys in the trees—he found tree kangaroos, which are marsupials (whose young develop in a special pouch). Why would such different animals be found in such similar habitats?

He put his thoughts to paper again. Wallace pointed out that under the doctrine of special creation, one would expect to find similar animals in countries with similar climates, and dissimilar animals in countries with dissimilar climates. This is not at all what he saw.

Comparing Borneo (in the west) and New Guinea (in the east), he wrote, "It would be difficult to point out two countries more exactly resembling each other in climate and physical features." But their birds and mammals were entirely different.

As he crossed the archipelago, Wallace noticed a striking pattern: mammals on the eastern islands resembled those of Australia and were marsupials, while those on the western islands resembled those of Asia and developed internally from a placenta. It was as if a line divided the archipelago (see Wallace's line in Figure 7.5).

Why would God draw a boundary through these islands and put monkeys in the trees on one side and kangaroos on the other? That made no sense to Wallace. Special creation could not explain the line, but Wallace's law that species come from preexisting species could. Wallace surmised that the eastern islands were once connected by land to New Guinea and Australia, and the western islands were never connected to the eastern ones, but were once connected to Asia. It was the history of the planet, not special creation, that explained the distribution of species.

GREAT MINDS THINK ALIKE

For Wallace the next burning question was not if species evolved, but how?

As he baked in a malarial fever on Halmahera in the Maluku Islands in early 1858, the answers came to him. (It has long been reported that Wallace was on the island of Ternate, and his report was posted from there. But recent research suggests that Wallace had traveled to Halmahera from Ternate.)

Alternating between hot and cold fits, Wallace had nothing to do but "to think over subjects then particularly interesting to me." Wrapped in a blanket on an 88 °F day, he thought of Malthus' essay on population, which he had read some years earlier. It occurred to him that the diseases, accidents, and famine that check the growth of human populations act on animals, too. He thought about breeding, how animals bred much more rapidly than humans, and if left unchecked, would overcrowd the world very quickly. But all of his experience revealed that animal populations were limited. Finding food and escaping danger ruled animal lives—and the weakest would be weeded out.

Wallace, the great collector, was also intimately familiar with the variety of individuals of a species. He realized that those that were better at finding food or evading enemies would outcompete those that were not.

Wallace wrote the paper out in full in just a few nights.

He entitled it "On the Tendency of Varieties to Depart Indefinitely from the Original Type." Wallace's paper was just a sketch, conceived in a dilapidated house during bouts of fever, 7500 miles from the center of science in England. Wallace did not send it directly to a journal; he wanted others to look at it first.

He sent it to a naturalist with whom he had begun a correspondence—Charles Darwin.

Darwin received Wallace's paper sometime in June 1858. He was shocked when he read it. Wallace had come to the same conclusions that Darwin had reached 20 years earlier, but had not yet published.

Neither author was aware of the other's thoughts and writing. How could two people with such different backgrounds come up with such similar, yet profound, ideas?

Great minds think alike.

Both men had seen nature up close and understood it was a battlefield. Both men had collected enough specimens of individual species to appreciate that species were variable. Both men had seen slightly different species restricted to particular islands and concluded that species change. Both men had read and recognized the relevance of Malthus' essay on population. The same patterns of facts led to very similar conclusions.

Wallace had asked Darwin to forward the manuscript to the geologist Sir Charles Lyell, which he did. Lyell and Joseph Hooker, the eminent botanist, were both close friends of Darwin's and knew he had formulated his theory of the origin of species by natural selection many years earlier. They decided that both men should share the credit. Lyell and Hooker took the initiative to arrange for Wallace's paper, and a brief summary from Darwin on his theory, to be read together at an upcoming meeting of the Linnean Society in London, and to be published together.

Neither Wallace nor Darwin was present at the meeting. Wallace was still in the Malay Archipelago and did not learn of the arrangement until after the fact. But neither the reading nor the publication of their papers drew much notice. It wasn't until the next year, when Darwin completed and published his book *On the Origin of Species*, that the public, the press, and the scientific world paid attention.

While still in the Malay Archipelago, Wallace received a copy of the book from Darwin. He read it over and over. Then, he shared his reactions in a private letter to his longtime friend Bates, who was back in England:

> I know not how or to whom to express my admiration of Darwin's book. . . .
> I do honestly believe that with however much patience I had worked up &
> experimented on the subject I could never have approached the completeness
> of the book,—its overwhelming argument, & its admirable tone & spirit. . . .
> Mr. Darwin has created a new science & a new Philosophy, & I believe that
> never has such a complete illustration of a branch of human knowledge, been
> due to the labours and researches of a single man.

It is perhaps the most generous compliment paid in the history of science.

Wallace returned to England in 1862, and the two men became lifelong friends. Wallace was always, for the rest of his life, deferential to Darwin. He always referred to the "Darwinian theory." He later dedicated his major book about his own epic journey, *The Malay Archipelago* (1869), to "Charles Darwin, author of the *Origin of Species*, not only as a token of personal esteem and friendship but also to express my deep admiration for his genius and his works."

END-OF-CHAPTER QUESTIONS

1. What is the doctrine of special creation? Why would a natural explanation of the origin of species be so radical that Darwin would suppress it for 20 years?

2. The year before Wallace sent Darwin his paper, Darwin was working on a book on his discoveries and ideas. On the left is an excerpt from that manuscript; on the right is a passage from Wallace. Neither man had access to the other's writing. What common experiences or observations could explain the similarities in language and ideas?

Darwin	Wallace
February to March 1857 Down, England	February 1858 Halmahera, Maluku Islands
Chapter 5. The **Struggle for Existence** as Bearing on Natural Selection	The life of wild animals is a **struggle for existence**. . . .
Any **variation**, however infinitely **slight**, if it did promote during any part of life even in the slightest degree, the welfare of the being, such variation would tend to be preserved or selected.	Perhaps all the **variations** . . . must have some definite effect, however **slight**, in the habits or capacities of the individuals. . . . A variety having slightly increased powers . . . must inevitably in time acquire a superiority in numbers.

3. The making of the theory of evolution is often told through Darwin's story, with little or sometimes no mention of Wallace. What would we miss about the scientific process if Wallace were left out of the story?

8

AN EXPLOSION OF ANIMALS

Only a small portion of the surface of the earth has been geologically explored, and no part with sufficient care.

—CHARLES DARWIN, *ON THE ORIGIN OF SPECIES* (1859)

ON THE AFTERNOON OF February 10, 1910, in Washington, D.C., three men in top hats and their best attire strode to the curb to board a sleek new horseless carriage. They paused for a moment that was captured by a photographer, three inventors whose deeds would become legendary and whose names would be linked forever with their revolutionary inventions—Wilbur and Orville Wright and Alexander Graham Bell (Figure 8.1).

Escorting these illustrious figures of American history out to their automobile was Charles Doolittle Walcott, a man whose name and deeds would be much less known to posterity, if at all. One might think that rubbing elbows that day with such famous men would certainly have been the highlight of Walcott's year, and perhaps of his whole life.

It was not even a close second.

Six months earlier, while riding his horse high in the Canadian Rockies over Burgess Pass, this nearly 60-year-old veteran geologist discovered one of the oldest and most important mother lodes of animal fossils ever unearthed. The remarkably well-preserved and somewhat strange animals of the Burgess Shale Formation mark one of the greatest chapters in the story of life, and one of the greatest mysteries in paleontology: the so-called Cambrian explosion—the

FIGURE 8.1 A Meeting of Legends On February 10, 1910, on the street in front of the Smithsonian Institution in Washington, D.C., Charles Walcott (left) escorted (from left to right) Wilbur Wright, Alexander Graham Bell, and Orville Wright to a waiting car.

Smithsonian Institution Archives. Image # 82-3350.

apparently sudden emergence of large, complex animals in the Cambrian period over 500 million years ago.

There was another reason why Walcott was unflustered by his eminent company that day. Walcott was quite used to circulating among the country's elite. Although he never finished high school or earned any kind of diploma or degree, Walcott rose to be the director of the U.S. Geological Survey, secretary of the Smithsonian Institution, a founder of the Carnegie Institution of Washington, and president of the National Academy of Sciences. By February 1910, he had already known and advised four presidents and would also serve the next three. And he carried out all of those duties while still managing to explore much of the geology of North America, and to make not one, but two landmark discoveries about the history of life.

His remarkable story and ascent began with exactly what would also mark his crowning discovery on that Canadian mountain peak—trilobites.

GROWING UP IN THE CAMBRIAN

Charles Walcott grew up in Utica, New York, and became a teenager in the middle of the U.S. Civil War. With many men away in the Union forces, there was a lot of work available for young Walcott. He started out working summers as a hand on William Rust's farm near Trenton Falls.

Dairying was the main source of income for Rust but he, like other farmers in the area, maintained a small quarry that was the source of the limestone for house and barn foundations, and some extra income. The area was well known for its Trenton limestone, which, in addition to being an abundant building material, was rich in fossils.

Walcott caught the fossiling bug early on. While quarrying was hard work, interrupted only by herding duties, it did provide the dividend of many fossils. Walcott learned where in the limestone layers the fossils were most abundant, and especially where to find the best trilobites, which were the most prized (Figure 8.2).

He also learned how different fossils characterized different layers of rock, and marked different geological eras. One day he found a block of sandstone

FIGURE 8.2 Trilobites These trilobites (*Ceraurus pleurexanthemus*) were collected by Walcott and sold to the Harvard Museum of Natural History.

on the road between Trenton and Trenton Falls. Walcott broke out all of the fossils, but none of them were of the Trenton limestone type he knew so well. He figured out that the trilobites must have been of pre-Trenton age, older than the trilobites he had been collecting up until then. Walcott deduced they must have been of Cambrian age—the earliest geological period then known to contain fossils. The name comes from the Latin word for Wales (*Cambria*). It was conceived by Rev. Adam Sedgwick after a series of field studies he conducted there in the 1830s, including an expedition to north Wales with a young assistant named Charles Darwin.

Walcott resolved right then and there that he would pursue the study of those older Cambrian rocks. That would not, however, be a formal course of study. His schooling ended at 17. Walcott found working for money and fossil collecting more appealing than school, and he had another motive for sticking around Rust's farm—he had his eye on Rust's daughter Lura. By the spring of 1870, when Walcott was 20, he was on the Rust farm full-time, and they were engaged within a year.

In between nursing sick cows, painting barns, baling hay, and setting up a house with his new bride, he was confident enough in his fossils and geology to contact some of the leading figures of academic paleontology and zoology. Walcott learned there was a market for his collections, and he needed the money in those hard times. Harvard paid $3,500—the equivalent of several years' wages—for his collection of 325 complete trilobites and crinoids, brachiopods, starfish, and corals.

Walcott later published his first paper in a scientific journal, the *Cincinnati Quarterly Journal of Science*, describing a new species of trilobite. He followed this with a series of other short reports, each growing a bit in technical detail and sophistication. These first successes were overshadowed, however, by his wife Lura's death in 1876, just four years after they were married.

Grief stricken, Walcott was at a loss until the chief geologist of the State Museum in nearby Albany offered him a job as an assistant. Walcott left his tidy trilobite farm in Trenton Falls and moved on to the state capital.

DOWN A GRAND STAIRCASE

It would be a relatively brief apprenticeship in Albany, just a bit under three years, but it was a valuable springboard and good therapy for the loss of his wife. Walcott learned more than geology from his boss, who was very well connected to politicians in the state legislature just down the street. The museum depended on their goodwill and support so Walcott made sure that the chief geologist looked good whenever an important visitor came by, a knack that would serve him well many years later. His boss reciprocated by recommending Walcott for a position with the fledgling U.S. Geological Survey (USGS).

On July 21, 1879, Walcott, then 29 years old, was appointed as a geological assistant, just the twentieth employee of the new USGS. He was assigned to a group whose task it was to map out the little-known geology of the Grand Canyon and surrounding regions. It had been just ten years since the one-armed Major John Wesley Powell and his companions had braved the rapids of the Colorado River and run the length of the Grand Canyon.

Walcott's specific assignment was to map out the geology of a long, nearly unbroken series of geological formations that rise from the Colorado River within the Grand Canyon up to the so-called Pink Cliffs in southern Utah, a gain of elevation of some 8000 feet. The cliffs, slopes, and terraces from the rim to the highest peak form what had been dubbed a "Grand Staircase," with each "riser" being a cliff or slope rising up to 2000 feet and separated by "treads"—plateaus up to 15 miles wide. Walcott was to measure his way down the staircase, starting from the summit of the Pink Cliffs and eventually reaching the Colorado River. He would identify each geological layer by the fossils they contained.

After a five-day train ride to Utah, and a 120-mile-long ride by stagecoach, Walcott set out from Beaver, Utah, by mule, traveling 10 to 15 miles a day. He was soon on his own, with just four animals and a cook. The USGS counted on the initiative of its geologists to figure out how to get the work done. Walcott was up to the task.

In a span of three months and using just a hand level, a length of chain, and an altimeter, Walcott measured off an 80-mile-long section more than 13,000 feet thick—a prodigious, unprecedented feat. He started in the rocks of the Pink Cliffs (now the Bryce Canyon area) and found fossils diagnostic of most major geologic periods as he descended—the Cretaceous, Jurassic, Triassic, Permian, Devonian—and, at the Colorado River, he found a few trilobites that marked the Cambrian. As the weather turned snowy, he wrapped up his work, loaded the mules, and carted his haul of more than 2500 fossils back to civilization. It was a six-day trek to a railway.

His superiors were more than a bit impressed. They doubled his salary and soon sent him west again. Walcott went to Nevada and the Eureka mining district where he concentrated on paleontology. Then, in the late summer of

1882, the new director of the USGS, John Wesley Powell himself, beckoned Walcott back to the Grand Canyon.

Because of Walcott's 1879 survey, the top of the canyon was fairly well described, but its depths were not. Powell, a solid geologist himself, wanted to know more about those magnificent formations he had seen as he rushed through the canyon years earlier in his wooden boat. He decided the best way to do that was to build a trail from the canyon rim down to the Colorado River. That would then give Walcott and others the opportunity to work during the winter months in the inner canyon, where it was warmer than up on the rim.

With an assistant fossil collector, a cook, and a mule packer under his supervision, Walcott began with a detailed study of the Tonto Group, a rock formation of Cambrian age. He then worked his way down, and down, and still farther down the geological sequence. Navigating the canyon was tricky as there was no continuous riverbank. Walcott and his team had to make trails as they went along the canyon walls and cliff sides, and up over ridges that connected one valley to the next. It was slow, dangerous going where one slip of a mule or man could mean a fall of several hundred feet. Adding to their woes and contrary to Powell's belief, winter did reach the inner canyon. The crew had to fight against winds and snow, and they melted chunks of ice by their campfires in order to give water to the animals. It became too much for Walcott's collector, who, too depressed by living in the depths of the canyon and the constant work, was sent home with Walcott's sympathy and blessing. After more than two months, the remaining crew and animals headed back up the trail out of the canyon and reached a camp on the rim with frostbitten feet.

Walcott had managed to document another 12,000 feet of section. When added to his 1879 study, it made for a "grand" total of 25,000 vertical feet surveyed—probably the largest section ever measured by a single geologist in the 19th century. But unlike his first trip, this one yielded very few fossils from the massive amount of rock he surveyed. His supervisor wrote to a British colleague about the mysterious emptiness of the fossil record in the rock below the Cambrian Tonto Group, "Anywhere Walcott cannot find fossils I pity anybody else who tries."

The absence of fossils below the Cambrian was a mystery not just of the Grand Canyon but of the whole planet. It was a mystery that confounded and troubled Darwin, but to which Walcott, though he did not at first realize it, held a few precious clues.

DARWIN'S DILEMMA

Darwin was well aware of the absence of fossils below the Cambrian. He candidly addressed the enigma in the *Origin of Species*. The vexing problem for Darwin and all other proponents of evolution was that trilobites and other creatures appeared suddenly in the Cambrian fossil record. It appeared as if the Cambrian explosion marked the dawn of life, and that complex animals emerged in a short span of time. This pattern in the fossil record did not fit with the theory of evolution, which predicted that complex organisms should evolve gradually and be preceded by simpler forms. Darwin admitted, "To the question why we do not find records of these vast primordial periods, I can give no satisfactory answer." Indeed, Darwin recognized that, if left unexplained, the void in the fossil record was a major problem for his theory.

In the Grand Canyon, below the lowest fossil-bearing layers of the Cambrian in which shelly creatures and trilobites were abundant, Walcott found a vast expanse of rock that was largely empty of life. He reported that he only found a few bits of fossils. Accustomed to finding hordes of specimens, he was understandably underwhelmed.

As it turned out, Walcott had been thrown off by a peculiarity of the rock formations in the Grand Canyon. In other places he had been in North America, the Cambrian was underlain by older, metamorphic rocks. In the Grand Canyon, however, he had found a vast section of typical-looking sedimentary rock below the lowest Cambrian, trilobite-bearing layer. He initially assumed that the several thousand feet of rock were, based on their appearance, also of Cambrian age. But Walcott was wrong. Several years later he realized that the section of sedimentary rock below the Cambrian fossils was in fact laid down during an immense period before the Cambrian—during the Precambrian era. The fossils he had found within it were the first clear evidence of Precambrian life and included structures now called stromatolites, formed by mats of ancient single-celled blue-green bacteria (cyanobacteria), and remains of large algae (Figure 8.3). Life did not begin all of a sudden in the Cambrian.

SNOWSHOE CHARLIE

It would be a very long time before Walcott would return to the Grand Canyon. There was a lot more geology to know, and over the next decade he went out on new expeditions every year. He roamed all over North America—to

FIGURE 8.3 Stromatolites from the Grand Canyon The layers were formed by mats of cyanobacteria and sediments. Walcott found these structures in the Grand Canyon in Precambrian rocks (700–1200 million years old), which was crucial evidence for early, simpler forms of life that preceded the emergence of animals. Photo by Carl Bowman, National Park Service.

Vermont, Texas, Utah, New York, Massachusetts, the Canada–Vermont border, North Carolina, Tennessee, Quebec, Colorado, Virginia, Alabama, Pennsylvania, Maryland, New Jersey, Montana, and Idaho—with almost all field work focused on the Cambrian.

A man with so much field experience was invaluable to the USGS, the leaders of which began to solicit Walcott's advice in guiding the survey's missions. Walcott became involved in explaining the survey's work and priorities to congressmen and senators on Capitol Hill. He always shot straight and showed such command of so many fields and issues that the politicians grew to trust and respect him.

In spring 1894, President Grover Cleveland nominated Walcott to become the new director of the USGS, and he was promptly confirmed by the Senate. With the survey so intimately involved in the exploration and management of the country's resources, he became a major figure of the scientific and governmental scene in Washington. He was adept at navigating federal agencies,

Congress, and even the White House. Walcott's quiet way of getting things done would earn him the nickname Snowshoe Charlie, for his adept maneuverings left no tracks.

It was just the sort of skill that presidents needed for sensitive topics. When President Theodore Roosevelt took on irrigation and water projects (dams, levees, etc.) as a major issue of his administration, he wanted them under the control of Walcott, who had been "tested and tried and we know how well they [the USGS] will do their work." Roosevelt's confidence in Walcott only grew as the director fended off special interests who wanted water projects for their states.

Roosevelt's and Walcott's passions converged most strongly on the matter of conservation. Walcott was frequently summoned to the White House for conferences about forests, rivers, parks, and politics. Concern that the United States lacked adequate laws to protect prehistoric artifacts, cliff dwellings, cemeteries, caves, mounds, and other sites, as well as areas of scientific or scenic value, helped forge the Antiquities Act of 1906. It empowered the president to set aside as a national monument, or park, whatever lands he deemed necessary for their preservation. One of the first areas Roosevelt designated as a national monument was the Grand Canyon (1908).

During his entire tenure as director of the USGS, and despite all of his other responsibilities, Walcott managed to escape Washington on a regular basis, particularly in the summers. After Walcott married his second wife, Helena, and once their children were old enough, these adventures came to be family affairs with the kids joining the treasure hunts. Nothing reinvigorated Walcott like the fresh air of the West, a good campfire, a ride on a sturdy horse, and—oh yes, a barrel of trilobites. When in Washington, he continued to work on his fossils in a side room off of his director's office.

At a time when most careers would be winding down, particularly one that had been so physically demanding, Walcott was still open to new challenges. In late 1906, the secretary of the Smithsonian Institution passed away and the top post was vacant. The regents wanted Walcott to take over. Roosevelt was reluctant to let him leave the USGS, but the president ultimately relented and the Smithsonian got "a man who was not only a recognized scientist, but a man of decided executive ability."

Roosevelt gave a White House dinner in Walcott's honor.

THE BIG BANG

Walcott's new job as head of the Smithsonian did not change his work habits. The summer field trips continued, and the Smithsonian staff soon learned that the secretary was not to be disturbed between ten in the morning and two in the afternoon, when he worked on his trilobites and other fossils.

In 1909, Walcott spent the second half of the summer in the Canadian Rockies, just as he had the two previous summers. He, Helena, and their 13-year-old son, Stuart, ventured to Alberta by train and eventually moved on to Field, British Columbia. From there, they rode their horses high up into the Yoho Valley. The dramatic scenery was often eclipsed by the equally dramatic weather, with thunderstorms, snow squalls, and sleet often forcing them to take shelter.

In the last days of August, they made their way up Burgess Pass and Walcott, Helena, and Stuart dismounted to go collecting. As they inspected some loose blocks of shale that had been brought down by a snow slide, they saw beautifully preserved crustaceans, and a lot of them. Walcott had never seen such fine specimens, nor had he encountered the species before (Figure 8.4). They hauled many samples back to camp. Luckily, the good weather held another day and they returned to the site to find equally fine, well-preserved sponges inside the slabs. The fossils were so remarkable, Walcott was eager to find the source of the fallen rock. After scampering up the slope, he did find more fossil-bearing slabs, but he could not find the original beds from which the slabs had fallen before the weather turned very nasty. Walcott knew he was on to something extraordinary, but his gang had to retreat for the season.

The next summer, in July 1910, Walcott, Helena, and two sons, Stuart and Sidney, eagerly headed back up the Yoho Valley. Once they established a camp, they worked their way up the 1200-foot slope above camp to start collecting. They quickly found a large number of what Walcott called "lace crabs," crustaceans less than an inch long that he would dub *Marrella* in honor of a colleague (Figure 8.5). They found gobs of trilobites, too, but they also found a lot of what Walcott described as "odds and ends," which were code words for creatures that baffled his expert eye.

They looked through every layer of limestone and shale to find the fossil-bearing band. When at last they found it, a vein about 7–8 feet thick and about 200 feet long, the operation went into high gear.

For 30 days they quarried the shale. Walcott had learned how to use dynamite when on Rust's farm and it was a handy way to remove a lot of rock from Burgess Pass. He and the boys slid the blasted blocks down the slope to

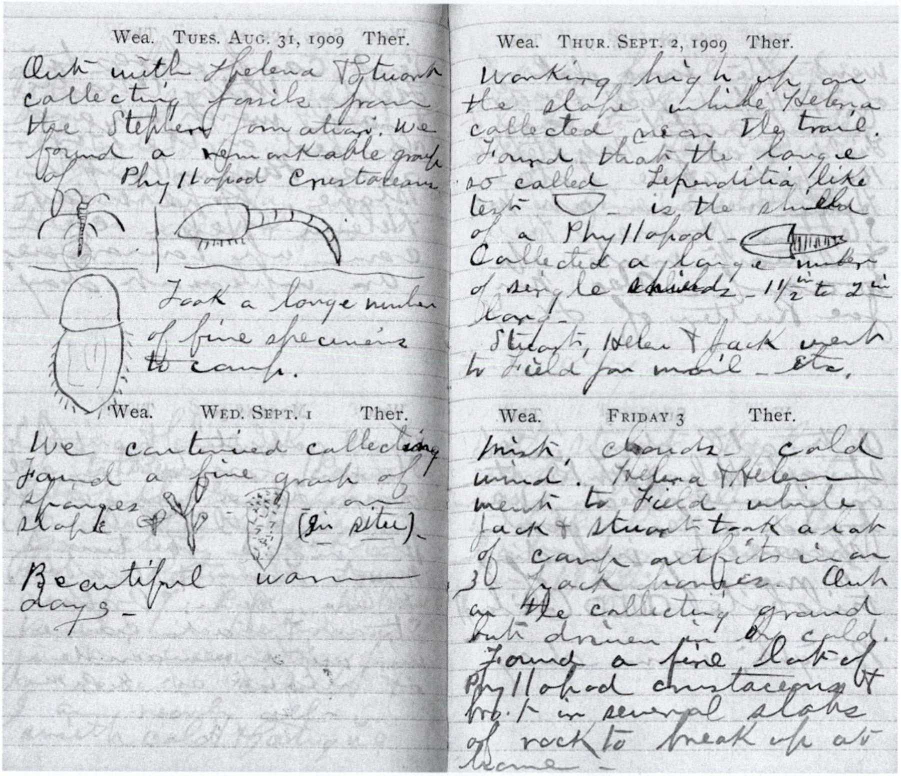
Wea. TUES. AUG. 31, 1909 Ther.
Out with Helena & Stuart collecting fossils from the Stephen formation. We found a remarkable group of Phyllopod Crustaceans. Took a large number of fine specimens to camp.

Wea. WED. SEPT. 1 Ther.
We continued collecting. Found a fine group of sponges on slope (in situ). Beautiful warm day.

Wea. THUR. SEPT. 2, 1909 Ther.
Working high up on the slope while Helena collected near the trail. Found that the large so called Lependitia like test is the shield of a Phyllopod. Collected a large number of single shields 1½ in to 2 in long. Stuart, Helen & Jack went to Field for mail etc.

Wea. FRIDAY 3 Ther.
Mist, clouds, cold wind. Helena & Helen went to Field while Jack & Stuart took a lot of camp outfits in on 3 pack horses. Out on the collecting ground but driven in by cold. Found a fine lot of Phyllopod crustaceans & brot in several slabs of rock to break up at home.

FIGURE 8.4 Days of Discovery The pages of Walcott's field diary containing first sketches of the Burgess Shale fossils. The drawings on August 31, 1909, depict a couple of crustaceans and a trilobite, and those on September 1, a sponge.

Smithsonian Institution Archives. Image # SIA2008-5399.

the trail where they loaded them onto pack horses that carried them down to camp. There, Helena led the operation of splitting the shale, trimming fossil-containing rocks, and packing them for transport to the railway station at Field, some 3000 feet below the camp. The operation flowed with the weather: on good days they blasted and hauled; on stormy days they stayed in camp and enjoyed splitting the shale and discovering the critters inside.

After a month, Walcott was exhausted, but elated. He wrote to a colleague at the Smithsonian, "The collection is great. . . . It has more fine, new things than I should have supposed could be found."

What he'd found was vivid evidence of a much more diverse animal kingdom in the Cambrian than had ever been seen before, or imagined. The Burgess Shale contained much more than the trilobites and brachiopods of other Cambrian deposits. The exquisite quality of the shale also preserved soft-bodied

FIGURE 8.5 Creatures of the Cambrian Animals of the Burgess Shale collected by Walcott: *Marrella* is a crustacean; *Sidneyia* is a large arthropod; *Ottoia* is a soft-bodied member of another phylum, the priapulids; and *Wiwaxia* has only recently been determined to be a mollusk. The fossils shown are about 4 to 8 cm. Courtesy of the Smithsonian Institution. Photos by C. Clark.

animals (without shells or hard outer skeletons) that some, including Darwin, thought could not be preserved in rocks. The creatures squashed within the dark Burgess Shale included representatives of many other major divisions of the animal kingdom (phyla), including annelid worms, priapulids, lobopodians, and even a chordate, a member of the phylum that includes humans (see Figure 8.5). The arthropods, a well-established Cambrian group thanks to the trilobites, were the most numerous animals in the shale, but these, too, were remarkable for the diversity of forms represented. Walcott gave colorful names to many of the creatures. One arthropod, for example, was dubbed *Sidneyia inexpectans* (Sidney's discovery) in honor of 14-year-old son Sidney's discovery of the specimen that defined the species.

The diversity of animal forms was both stunning and baffling. Bizarre creatures such as the five-eyed *Opabinia* and the largest fossil, *Anomalocaris*, were also named by Walcott, who took his best shot at classifying them, assigning them to new arthropod orders or families within existing classes. Subsequent research has revealed them to be related to, but different from, any living classes of arthropods. Other Burgess animals, such as *Wiwaxia*, long defied firm classification.

Regardless of their exact taxonomy, the significance of these fossils is manifold. The Burgess Shale offered the first glimpse into a riot of life-forms in the Cambrian seas. Nothing like it had been found before, and few locales have since been found that can rival it. For many of the soft-bodied animals, the Burgess Shale was not just their earliest appearance in the fossil record, but their only fossil record at the time. Their discovery pushed the origins of most modern animal phyla back in time to at least the Cambrian. With so many phyla making their first appearance in the Cambrian, later scientists came to dub this period the Big Bang of animal evolution.

A century after Walcott's discovery, we know quite a bit more, but certainly not all we would like to, about the Cambrian explosion. The most solid set of facts we have concern the time frame of the explosion, which was completely beyond the reach of methods available in Walcott's day (the age of the earth had not been settled either and estimates varied widely). Walcott thought that the fossils were 15 to 20 million years old. We now know that they are much, much older—the Cambrian period began 543 ± 1 million years ago—and that the Burgess fossils are about 505 million years old. The Precambrian rocks in which Walcott found a few fossils were formed 700 million to 1.2 billion years ago.

Other Cambrian deposits have since been discovered that are older than the Burgess Shale, and that also contain a great variety of animal forms. They reveal that most animal groups began emerging by about 530 million years ago, and thus the explosion was underway for some 20–25 million years before the Burgess animals were buried and fossilized.

What lit the fuse that ignited the explosion? Why was life generally small and simple for so long, and why and how did large, complex forms seemingly burst on the scene? Geochemists have determined that there was a dramatic rise in oxygen levels in the late Precambrian seas. Since larger creatures need to distribute oxygen to cells deep within their bodies, many scientists favor the idea that rising ocean oxygen levels made possible the emergence of larger animals. The evolution of more and larger animals may then have triggered an explosive ecological competition among predators and prey that rapidly filled the seas with all sorts of creatures.

LEGACY

Walcott returned to the Burgess quarry for several more seasons after 1910 and to the Canadian Rockies every year through 1925, until he was 75. He managed to bring back an astonishing hoard of more than 65,000 Burgess specimens to the Smithsonian Institution, which are now a national treasure.

Walcott did not neglect other collections or missions of the national museum during his 20-year-long administration. Walcott was later pivotal in getting Congress, the military, and President Wilson to establish in 1915 a National Advisory Committee for Aeronautics (NACA), whose executive committee he wound up chairing. Forty-three years later, in 1958, as the country rushed to catch up to the Soviet Union after the launch of *Sputnik*, the organization Walcott helped to found was morphed into NASA and given the charge to explore outer space.

From the depths of the Grand Canyon to the top of the Canadian Rockies, from the microscopic relics of the Precambrian to the animals of the Big Bang, from the opening of the American West to the opening of the space race, from Trenton Falls to the White House, Snowshoe Charlie left some very deep tracks.

END-OF-CHAPTER QUESTIONS

1. Why was Darwin so concerned about the seemingly sudden appearance of animals in the Cambrian?
2. Why was Walcott's discovery of fossils in Precambrian rocks of the Grand Canyon important?
3. Walcott sought out and explored Cambrian rocks all over North America over many years. Why are the fossils of the Burgess Shale significant?
4. What characteristics did Walcott demonstrate that you think might be important in becoming a scientist?
5. Why might the animals of the Burgess Shale be difficult to classify with respect to modern animals?
6. The Burgess Shale was formed about 505 million years ago. The age of the earth is ~4.5 billion years, and the earliest known evidence of life is ~3.5 billion years ago. But for 3 billion years, most life on earth was microbial.

a. What might be one explanation for why it took so long for larger creatures such as animals to appear in the Cambrian?

b. What might be an explanation for the rapid appearance of so many different kinds of animals in the Cambrian?

9

EVOLUTION BY MERGER

Any living cell carries with it the experiences of a billion
years of experimentation by its ancestors. You cannot expect
to explain so wise an old bird in a few simple words.
—MAX DELBRÜCK, *A PHYSICIST LOOKS AT BIOLOGY* (1949)

SHE WAS COMING UP the stairs of Eckhart Hall, the math building at the University of Chicago. He was coming down. They caught each other's eye, and stopped to chat.

Her name was Lynn. She was a freshman from a tough neighborhood on the south side of Chicago where the schoolgirls carried razor blades. Eager to get away from home, precocious, and self-confident, she had enrolled at 14 in a special program at the university's high school to take more challenging classes. She entered college at just 16.

His name was Carl. He was a 20-year-old physics major who grew up in New Jersey, who had also entered the prestigious and rigorous university at age 16. She thought he was tall and handsome, and he seemed unusually articulate—he was certainly full of conversation. Carl was president of the university astronomy club, wrote articles about space, and gave public talks. Lynn got the impression that Carl was somewhat "famous" on campus.

They started dating, and talking. They both loved to talk—a lot.

Ever since he was a young boy, Carl had been gripped by the prospect of life elsewhere in the solar system and the universe. Now, in 1954, at the dawn of molecular biology, some of the secrets of life on earth were beginning to

111

be revealed. Lynn knew very little about science—she described herself as an "ignoramus." She was hoping to become a writer.

Carl, on the other hand, was swimming in science, full of ideas. He loved to ramble on about space travel, and although nothing had yet been launched into space, he bet his friends that humans would land on the moon by 1970. Carl took it on himself to introduce Lynn to the latest science, including what she thought were some far-fetched notions about life on other planets.

Lynn was great company: she was interested in ideas, completely open-minded, easily excited, and had a great sense of humor. Her warmth and sharp intellect won many friends. And she loved Carl, his infectious enthusiasm, and warm smile.

Inspired by Carl, Lynn's inner scientist awoke. A yearlong natural sciences course provided the opportunity for Lynn to read the classic works of many pioneering biologists, including Darwin (Chapter 7) and Morgan (Chapter 3). Their discoveries and ideas drew her toward genetics and the conviction that understanding heredity was the key to evolution. As Lynn earned her bachelor's degree and headed toward a future in science, Carl stayed at Chicago to pursue a second bachelor's degree, and then a master's degree in physics before entering the PhD program in astronomy.

But with Carl's brilliance and ambition came a healthy dose of self-centeredness that often infuriated Lynn. She broke off their relationship several times over her undergraduate years, only to be wooed back by Carl's considerable persuasion and charm. Despite the turbulence, and Lynn's misgivings, they started talking marriage. Even though she was fiercely independent and determined to pursue her own scientific career, this was the 1950s. "You did not go with a man and live with him or sleep with him in some public way . . . unless you married first," she later explained.

So, on June 16, 1957, a week after receiving her bachelor's degree, in an uncomfortably hot Chicago hotel, 19-year-old Lynn Alexander married 22-year-old Carl Sagan (Figure 9.1).

The two had very bright futures although, as it would turn out, not with each other. Carl Sagan would become a world-famous astronomer and public figure. But it would be Lynn, not Carl, who would make a major contribution to the story of life.

FIGURE 9.1 Lynn Alexander and Carl Sagan Marry June 16, 1957, Chicago, Illinois.
Photo courtesy of the estate of Lynn Margulis.

TOO FANTASTIC FOR POLITE SOCIETY

Carl's doctoral work was to be based largely at the Yerkes Observatory in Williams Bay, Wisconsin. Lynn applied to and was accepted into a master's program in genetics and zoology at the University of Wisconsin, and the couple moved to Madison.

Carl supported Lynn's career ambitions, but he also envisioned a large family of brilliant children. He planned it out on a timetable: if they had one child every 18 months, by the time she had five, Lynn would then be able to focus on training and her PhD.

Lynn was surrounded by exceptional mentors and teachers at Wisconsin. Her research advisor, Walter Plaut, was a very warm and enthusiastic person who introduced Lynn to the inner world of amoebae. When she peered through the phase-contrast microscope, it was love at first sight. Lynn enjoyed feeding the colorful protozoans—siphoning out the dirty water from the culture bowl and adding food and salts—even when that required going to the lab on Saturday nights.

It was her cytology professor Hans Ris who sparked the mental rioting that would become her life's work. In his advanced class, Ris had the habit of reading aloud from E. B. Wilson's classic 1925 book, *The Cell in Development and Heredity*. In it, Morgan's close colleague at Columbia (see Chapter 3) described the ideas of a scientist named Wallin, who in 1922 had suggested that mitochondria "may be regarded as symbiotic bacteria whose associations with other cytoplasmic components may have arisen in the earliest stages of evolution." Lynn was electrified—mitochondria evolved as symbiotic bacteria?

Ris continued reading, "To many, no doubt, such consideration may appear too fantastic to mention in polite scientific society; nevertheless, it is in the range of possibility that they may some day call for more serious consideration. . . ."

Symbiosis, one organism living in close association with another, was known in zoology but was considered a rare, special case. The idea that organelles within complex eukaryotic cells arose from bacteria living inside a host cell (endosymbiosis) was truly radical. That was not the origin of species as Darwin conceived it, as the splitting of one lineage into two, but a joining of two into one. Nor was it genetics as Mendel and Morgan described, for the genetic action would all be in the cytoplasm and follow different rules than genes in the nucleus. Moreover, at the time, bacteria were largely thought of as pathogens that caused disease in animals and plants, not a source of key structures.

Polite or not, Lynn was gripped by the possibilities and the implications. She had read about the inheritance of certain traits in various algae that did not occur in typical Mendelian fashion through nuclear genes, but was dependent on factors in the cytoplasm. Did mitochondria or chloroplasts carry their own genetic information? And if so, where did they get it?

As it so happened, Madison was a good place to begin her investigations as both Plaut and Ris were also keenly interested in organelles. During her first year in the lab, she and Plaut fed a radioactive precursor of DNA (thymidine) to amoebae growing in culture. Then, they traced where the radioactivity accumulated by fixing the cells on a slide, placing them in the dark, and overlaying a silver-containing photographic emulsion. After waiting 77 days (!) for the emulsion to be exposed, they looked in the microscope at where the silver grains—the indicator of radioactive thymidine—were visible.

The thymidine accumulated in the cell nuclei in DNA, as expected. But it also appeared all over the cytoplasm. They showed that this was due to incorporation of thymidine into cytoplasmic DNA, because treatment with a DNA-destroying enzyme eliminated the radioactive signal. This was strong evidence that the synthesis of DNA also occurred in the cytoplasm, but for what purpose they could not say.

Lynn's time in Madison was gratifying in other ways. Right on schedule, she gave birth to their first son (Dorion) in 1959, and was pregnant with their second (Jeremy) in 1960 when she completed her master's degree. Carl received a prestigious fellowship to work at the University of California, Berkeley, so Lynn applied to and was admitted into the doctoral program in genetics. The family pulled up stakes and moved west.

APARTHEID

As Carl hurled himself into several projects, including the *Mariner 2* mission to Venus and a project to send a balloon into the stratosphere, Lynn sought to continue her exploration of the possible symbiotic origin of organelles.

At the time, a fundamental division was recognized between prokaryotes—unicellular organisms such as bacteria and cyanobacteria that lacked cell nuclei, mitochondria, or chloroplasts, and did not divide by mitosis—and eukaryotes, whose cells contained cell nuclei, mitochondria, or chloroplasts (in algae and plants) and divided by mitosis. The "pro" in prokaryote implied that these simpler cells were the evolutionary precursor to the more complex eukaryotic cells. However, the differences between the two forms of cells appeared as the greatest discontinuity in all of evolution. While it was possible to conceive, for example, how fish with fins may have gradually given rise to four-legged descendants, it was difficult to envision how prokaryotes gradually accumulated all of the structures and machinery characteristic of eukaryotes.

Lynn realized she needed a very broad background to tackle the puzzle, more than just genetics and zoology, but grounding as well in the evolutionary history of life, a field dominated by paleontology. To her dismay, she discovered that there was very little discussion of evolution in the genetics department, very little knowledge of genetics among bacteriologists, and no contact between the department of paleontology, where evolution was studied, and other groups. She was shocked at the extent of what she called "academic apartheid."

She would have to seek the connecting clues on her own. Much of that search took place not in the laboratory, but in the library. She collected evidence of cytoplasmic (nonnuclear) inheritance in various species of plants and algae. She read about mutants in yeast (a fungus) that were deficient in mitochondrial respiration that provided more evidence that organelles might contain their own genes. And she delighted in beautiful microscopic studies from her former mentors Plaut and Ris.

One of the leading microscopists in the world, Ris was expert in the use of a high-powered electron microscope that revealed cellular architecture in unprecedented detail. In 1962, he and Plaut trained the microscope on *Chlamydomonas moewusii*, a green alga, and revealed DNA-containing structures within its chloroplasts. They also noted similarities in the organization of chloroplasts and cyanobacteria (a prokaryote) in the way membranes enclosed their photosynthetic machinery, and the presence of protein-synthesizing ribosomes that were associated with the organelles. The authors wrote:

> We suggest that this similarity in organization is not fortuitous but shows some historical relationship. . . . With the demonstration of ultrastructural similarity of a cell organelle and free living organisms, endosymbiosis must again be considered seriously as a possible evolutionary step in the origin of complex cell systems.

The next year, a Swedish team found similar microscopic evidence of DNA in mitochondria. No one could say where these organelles' DNA came from, but the prospects for endosymbiotic origins were improving. These encouraging discoveries, however, were tempered for Lynn by domestic strife.

Carl had overcommitted himself to too many projects, and traveled frequently. And when he was home, he did not pitch in at all with child-rearing, or anything else. Lynn took care of everything from meals to dog walking, while raising two children and managing her doctoral work. She was fed up and exhausted. As conflict escalated, Lynn decided to end the marriage. Carl tried to talk her out of it, but his considerable powers of persuasion failed him this time.

SYNTHESIS

Carl had received an offer from Harvard, so even though she had not yet received her PhD, Lynn moved to Boston so that the kids could be near their father. To continue her research, she worked in a lab at Brandeis University. At Berkeley, she had tried to localize DNA in *Euglena gracilis*, another single-celled photosynthetic eukaryote, using the same radioactive thymidine technique that she and Plaut had employed at Wisconsin. That did not work. At Brandeis, Lynn discovered that a different base, guanine, was readily incorporated into nuclear and cytoplasmic DNA. This finding added to a growing body of evidence for DNA in chloroplasts.

Lynn's more immediate concern was making ends meet, so she took a part-time position at Boston University lecturing. She was in a fragile position professionally. Her experimental work was sufficient to earn her a PhD, but it was neither that original nor powerful with respect to the question of organelle origins. Her appointment was not on the professorial tenure track, and she was a single mother with primary responsibility for two young children.

Nonetheless, the question of the origin of eukaryotes was a huge mystery, and she remained determined to solve it. She decided to try to assemble all of the previous thought and available evidence concerning endosymbiosis into one synthesis. She was keenly aware that the hypothesis of the endosymbiotic origins of eukaryotic organelles was not new—various authors had put forth or supported the idea over several decades. But no one had put together all of the new evidence in light of what was known about the history of life and the earth, or suggested how the idea might be further tested.

Lynn developed a scheme in which a series of endosymbiotic events gave rise to eukaryotes. New fossil discoveries reported in 1966 had stretched the then-known history of life back to about 3.1 billion years ago, and the first appearance of eukaryotes to about 800 million years ago (the history of life has since been extended to about 3.5–3.8 billion years ago, and the first appearance of eukaryotes is now placed at about 2 billion years ago). Lynn underscored the long, prokaryote-dominated history of life that preceded the rise of eukaryotes. She proposed that the first endosymbiotic event was the acquisition of the protomitochondrion, which was followed later by the acquisition of a photosynthetic prokaryote that eventually became the chloroplast in algae and plants (Figure 9.2).

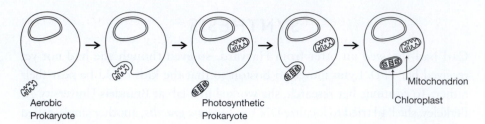

FIGURE 9.2 The Endosymbiotic Origin of Eukaryotes The basic scheme proposed by Lynn Sagan (Margulis) entails the acquisition by a host cell of a bacterial symbiont that became the mitochondria of eukaryotes. Subsequently, the acquisition of a photosynthetic cyanobacterial symbiont led to the evolution of chloroplasts in algae and plants.

Lynn relied heavily on the recent cytological studies that revealed:

i. Both mitochondria and chloroplasts contain specific DNA and RNA;
ii. Both organelles are self-reproducing structures, and each organelle arises from a preexisting organelle and is not assembled from scratch;
iii. Each organelle contains a genetic system that is inherited separately from the nucleus and is responsible, at least in part, for organelle function.

On the basis of this evidence, Lynn suggested that endosymbionts originated as free-living cells that were once able to replicate their own DNA and synthesize their own proteins. She proposed that eukaryotic cells should be seen as "multi-genomed" systems, an amalgam of multiple organismal histories. Moreover, she predicted that organelle DNA contained genetic information from these endosymbionts, such that, for example, genes might be found in common between cyanobacteria and chloroplast DNA.

Lynn submitted the manuscript. It was rejected. She submitted it again. It was rejected again . . . and again . . . and again. Finally, after about 15 rejections, the 50-page paper was accepted by the *Journal of Theoretical Biology* and published in 1967.

Resistance to the theory came from at least two directions. Some biologists were disturbed by the idea that such important structures in evolutionary history could have been inherited wholesale through a merger of different lineages, rather than evolving gradually in one lineage through many intermediate steps as the established Darwinian theory of evolution prescribed. The second line of objections concerned the strength of the evidence available. Such an extraordinary claim required powerful evidence, and some biologists were not persuaded that the similarities seen excluded the alternative that eukaryotes and their organelles evolved stepwise from prokaryote ancestors.

Regardless of these objections, Lynn's paper on the theory of endosymbiosis garnered a lot of attention, and secured her a tenure-track position at Boston University. She also remarried in 1967 to crystallographer Thomas Margulis, and changed her last name from Sagan to Margulis.

When Lynn became pregnant in 1969 with her daughter Jennifer, she had to stay home for health reasons. This enforced leave allowed her first long period of uninterrupted thought. She used the time to expand her 1967 paper into a heavily illustrated book entitled *Origin of Eukaryotic Cells*. After many late nights typing to meet a contract deadline, Lynn carefully packed the manuscript and artwork and sent it to her publisher.

Then she waited. And waited. She received no acknowledgment for five months. Finally, the box came back to her with the manuscript inside, with no letter or explanation. Lynn eventually found out that extremely negative reviews had persuaded the publisher to abandon the project. After spending another year revising the book, it was finally published—by a different publisher.

FROM REBEL TO RESPECTED

The overturning of established ideas, and the acceptance of a new scientific theory, are part of the scientific process. Surmounting doubts, both reasonable and unreasonable, however, can be a long journey. The widespread acceptance of the theory of evolution (Chapter 7), or even just the infectious origin of ulcers (Chapter 1), required multiple lines of evidence. They also required time—lots of time—for doubts, or doubters, to vanish.

This was also the case for the theory of endosymbiosis. Even though endosymbiosis was not her original idea, and most of the supporting evidence had been obtained by others, Lynn found herself the chief explainer and defender of the theory. She understood the major sources of resistance. The endosymbiotic theory was a revolutionary departure from the Darwinian process of species formation. Moreover, the events had taken place so early in the history of life, under environmental conditions and among organisms that no longer existed.

Lynn acknowledged the challenge.

> It is never possible to prove rigorously that a unique series of events did occur in any historical context. Evolutionary biologists are in the same logical

predicament as historians: they deal with series of complex irreversible phenomena and can only present arguments based on the assumption that of all
the plausible historical sequences one is more likely to be a correct description
of the past events than another. . . . The probability that a particular historical reconstruction is more accurate than another increases markedly if the
number of present observables that can be explained increases.

It was difficult to see how the theory could be tested more rigorously. But,
in 1975, two groups of molecular biologists used new techniques to compare
the structures of an RNA molecule (so-called 16S rRNA) found in the ribosomes of bacteria and chloroplasts. They discovered close similarities between
the 16S rRNA of chloroplasts of two species of algae and a ribosomal RNA
in photosynthetic cyanobacteria. Two years later, close similarities were found
between a specific type of mitochondrial ribosomal RNA and bacterial ribosomal RNA (see Chapter 10).

Individually, these discoveries were strong evidence that chloroplasts and
mitochondria each evolved from bacterial symbionts. Together, they were even
more powerful evidence for endosymbiosis because each organelle appeared
to be more related to cyanobacteria and proteobacteria, respectively, than to
other bacteria and thus to have arisen from different symbionts. This observation refuted the objection that the organelle genes could have simply evolved
stepwise from nuclear genes, because that ancestor could not be two different
lineages of bacteria at once (without endosymbiosis).

Eukaryotes, and algae and plants in particular (with their chloroplasts),
were, as Lynn had asserted, multigenomed systems. With this fresh molecular evidence, the endosymbiotic theory rapidly gained acceptance, and Lynn
gained recognition for being its champion and explainer. She also never abandoned her aspirations of becoming a writer, for she subsequently wrote many
expert and popular books on symbiosis and evolution.

In a ceremony at the White House in 2000, Lynn Margulis was awarded the
National Medal of Science: "For her outstanding contributions to understanding of the development, structure, and evolution of living things, for inspiring
new research in the biological, climatological, geological and planetary sciences, and for her extraordinary abilities as a teacher and communicator of
science to the public" (Figure 9.3).

FIGURE 9.3 Lynn Margulis Receives the National Medal of Science from President Bill Clinton The ceremony occurred on March 14, 2000, in the East Room of the White House.

Paul Hosefros/The New York Times/Redux.

END-OF-CHAPTER QUESTIONS

1. Why did earlier biologists consider the symbiotic origin of eukaryotic organelles such a radical idea—too fantastic for polite society?
2. In what way is symbiosis a simpler or easier explanation for the origin of eukaryotic organelles than a gradual Darwinian explanation?

3. What new pieces of evidence prompted Lynn and other biologists to resurrect symbiotic explanations of organelle origins?

4. What did Lynn mean by the description of eukaryotes as multigenomed systems? What evidence emerged that this was indeed the case?

10

A THIRD FORM OF LIFE

The time will come, I believe, though I shall not live to see it, when we shall have fairly true genealogical trees of each great Kingdom of Nature.

—CHARLES DARWIN, LETTER TO T. H. HUXLEY, SEPTEMBER 26, 1857

HE WASN'T LOOKING FOR a new kingdom.

One day late in the summer of 1966, Indiana University microbiologist Tom Brock and his student Hudson Freeze were prowling around the geysers and hot springs of Yellowstone National Park. They were interested in finding out what kinds of microbes lived around the pools and were drawn to the orange mats that colored the outflows of several springs.

They collected samples of microbes from Mushroom Spring, a large pool in the Lower Geyser Basin whose source was exactly 163 °F (73 °C), thought at the time to be the upper temperature limit for life (Figure 10.1). They were able to isolate and culture an organism from their sample, a new species that thrived in hot water. In fact, its optimal growth temperature was right around that of the hot spring. Brock also noticed some pink filaments around some even hotter springs, which raised his suspicion that life might occur at even higher temperatures.

The next year, Brock tried a new approach to "fishing" for microbes in the hot springs of Yellowstone. His fishing tackle was simple: he tied one or two microscope slides to a piece of string, dropped them in the pool, and tied the other end to a log or a rock (don't try this on your own—you will be arrested and quite likely scalded or worse). Days later, on retrieving the slides, he

FIGURE 10.1 T. D. Brock at Yellowstone Thermal Pool

Photo courtesy of Tom Brock, Madison, Wisconsin.

could see heavy growth, sometimes so much that the slides had a visible film. Brock was right that organisms were living at higher temperatures than had previously been thought, but he did not imagine that they were living in *boiling water*. And they weren't just tolerating 200 °F or more—these organisms were thriving in smoky, acidic, boiling pots such as Sulphur Caldron, in the Mud Volcano area of the park.

As the discoverer of these remarkable microbes, Brock had the privilege of naming them. He chose memorable Latin monikers that reflected the extreme habitats in which they were found: he dubbed his first hot springs discovery *Thermus aquaticus*; a subsequent sulfur-spring find *Sulfolobus acidocaldarius*; and a third species from a hot Indiana coal pile *Thermoplasma acidophila*. Brock also had the responsibility for classifying them. Under the microscope, each of Brock's new species had the size, shape, and overall appearance of various bacteria—being long filaments, shorter rods, or spheres (Figure 10.2). The question was, to which group of bacteria did they belong?

Brock determined that their physiological characteristics were sufficiently distinct that they did not belong to any existing genus (a grouping one rank above species in the conventional Linnaean classification scheme), and boldly decided that they each deserved their own genus.

FIGURE 10.2 A Sample of Microbes from a Hot Spring This scanning electron micrograph reveals the growth of a variety of microbes on a slide that was immersed in the Obsidian Pool of Yellowstone National Park.

Hugenholtz et al. (1998). © 1998, reproduced with permission of the American Society for Microbiology.

Brock's amazing finds greatly expanded biologists' views about the conditions that life could tolerate. They also fanned speculation that these so-called thermophiles found in hot springs might be relics of primordial forms of life that evolved in the hotter environments of the early earth, and later gave rise to more familiar bacteria.

But this latter idea was nothing more than speculation because the evolutionary relationships among bacteria were very poorly known.

At the time, most classification was based on morphology. However, unlike animals and plants that possessed many visible characteristics that defined larger groups (e.g., fish, birds, reptiles, mammals), bacteria had a limited range of physical forms such that unrelated species could bear a similar appearance. As a result bacterial genealogy was a mess. "The ultimate scientific goal of biological classification cannot be achieved in the case of bacteria," stated the leading textbook at the time.

The task seemed so difficult that most microbiologists essentially threw up their hands and gave up.

But Carl Woese was not most microbiologists.

THE RIGHT MOLECULE, AND THE RIGHT MAN, FOR THE JOB

A professor at the University of Illinois, Woese was keenly interested in the evolution of cellular life. He knew that bacteria were key to understanding life's early history, but because of the sorry state of bacterial genealogy, neither he nor anyone else could say which groups might offer important insights. Woese was determined to break the impasse and thought that he just might see a new way to do so.

Over the course of the 1960s, Woese had struck up a friendship and correspondence with Francis Crick. In June 1969, Woese divulged his plan to Crick:

> Dear Francis:
>
> I'm about to make what for me is an important and nearly irreversible decision, and would be grateful to you for any thoughts you have on the matter and for any backing (largely moral) you could give me.
>
> If we are ever to unravel the course of events leading to the evolution of the prokaryotic (i.e., simplest) cells, I feel it will be necessary to extend our knowledge of evolution backward in time by a billion years or so—i.e., backward into the period of actual "Cellular Evolution." There is a possibility, though not a certainty, that this can be done by using the cell's "internal fossil record"—i.e. the primary structures of various genes.

Woese had spent several years theorizing about the evolution of the genetic code and the process of translation. Because the code was shared among all organisms, Woese reasoned that the translational machinery of the ribosome must be ancient. He continued:

> The obvious choice of molecules here lies in the components of the translation apparatus. . . . I feel (and you may too) that the RNA components of the machine hold more promise than . . . protein components.

Two different RNA molecules, 5S ribosomal RNA and 16S ribosomal RNA, are integral parts of the ribosome ("S" is a unit of measurement that differs with the size of molecules). Woese hoped that the sequences of these components might retain a record of evolutionary relationships.

Crick, a theorist who had not conducted an experiment in perhaps 20 years, replied, "I think the project is a good one and an important one but it may

FIGURE 10.3 **Carl Woese and RNA Fingerprints** Carl Woese in front of a light box with RNA fingerprints to be analyzed.

Photograph courtesy of the University of Illinois at Urbana-Champaign Archives/Carl Woese Papers, Record Series number 15/15/22 box 19.

well be difficult to get money for it, especially as it's rather a gamble whether enough evidence is still frozen in the sequences."

While no technology existed at the time to determine the entire sequence of an RNA molecule, a method had been developed that fragmented RNA into many short pieces, which could in turn be separated into sets of spots in a two-dimensional pattern (dubbed a "fingerprint") (Figure 10.3). The sequences of individual spots (short RNA fragments) could then be determined using additional techniques. Each species produced a unique "catalog" of short RNA sequences.

Woese's plan was to assemble the RNA catalogs of a number of bacteria and eukaryotes, and then use the degree of similarities and differences within each catalog to determine evolutionary relationships. The general idea was that RNA sequences would evolve over time, so that more closely related organisms would have more similar catalogs than more distant relatives.

It took several years to establish and improve the necessary techniques. It was also a long process to determine the catalog of any one species, so Woese organized an assembly line approach. Every other Monday, the lab received a large quantity of radioactive phosphorus-32. This was then divided up and fed to cells in a low-phosphate medium so that newly synthesized RNA molecules would incorporate the ^{32}P. The 16S and 5S ribosomal RNAs were isolated, and then digested with an enzyme to create an initial fingerprint.

The analysis of the fingerprints was Woese's job. He spent almost every hour of every day in the lab, eight hours a day, five days a week, analyzing X-ray films that were held up against a light screen. Woese annotated the spots with Sharpie markers, and noted which ones would be isolated and sequenced. It was slow, painstaking work that demanded his full concentration. Some days he would walk home from the laboratory saying to himself, "Woese, you have destroyed your mind again today."

After the process was complete, which took about three to four weeks, the catalog of short RNA sequences from each species (oligonucleotides or "oligos") was entered into a master record for further comparative analysis. As the number of species catalogs grew, it was time to see whether the molecules indeed held clues to events in the deep past.

HISTORY ON A MOLECULE

As Woese analyzed more species, he became able to quickly recognize certain spots that he had seen before. The bacteria and eukaryotes were readily distinguishable by sets of oligos that were restricted to each group. For example, there were two spots that appeared in the fingerprints of all bacteria. There were also "universal" oligos that were found in both groups, as well as oligos that were limited to certain bacteria or eukaryotes.

Woese and his postdoctoral fellow George Fox interpreted the catalogs in evolutionary terms. For example, bacteria shared more oligos among themselves than with eukaryotes, reflecting that there was one distinct root to the bacterial evolutionary tree. Likewise, eukaryotes shared more oligos among themselves than with bacteria. The universal oligos in turn reflect the deep origin of ribosomal RNA (rRNA) in a common ancestor of bacteria and eukaryotes.

These patterns of shared RNA sequences were very gratifying. They raised the hope that the information in rRNA was rich enough to decipher relationships within each group, or to resolve even trickier questions such as the genealogy of eukaryotic organelles.

Well aware of Lynn Margulis' endosymbiosis proposal (Chapter 9), and the controversy it generated, Woese first set his sights on the question of the origin of chloroplasts. His team isolated chloroplasts from the single-celled photosynthetic eukaryote *Euglena gracilis*, the very species in which Margulis had detected chloroplast DNA. They next isolated 16S rRNA, digested it to create a fingerprint, and then worked out the sequences of the oligos in *Euglena*'s catalog. They also isolated the cytoplasmic (non-chloroplast) 18S rRNA from *Euglena*.

They found that the *Euglena* chloroplast 16S rRNA was strikingly related to bacterial rRNA, but that cytoplasmic *Euglena* rRNA was not. Numerous oligos were shared with bacterial species, including four oligos six bases or longer that were shared with a photosynthetic bacterium and no other species. The sequences shared between *Euglena* and bacterial rRNA provided strong evidence for a bacterial origin of chloroplasts and the endosymbiont theory.

The evolutionary connection between bacteria and chloroplasts was confirmed and extended in a parallel study by Linda Bonen, Woese's former technician, and W. Ford Doolittle at Dalhousie University in Halifax. Woese had encouraged Bonen to share the lab's RNA sequencing process with Doolittle, a good friend. Bonen and Doolittle deciphered the oligo catalog of the chloroplast rRNA from a red alga, and discovered that it too was very similar to bacterial rRNA, as well as to *Euglena* chloroplast rRNA.

Bonen and Doolittle then went on to team up with Michael Gray and Scott Cunningham, two colleagues at Dalhousie, to examine mitochondrial rRNA. They found that mitochondrial 18S rRNA from wheat was highly related to bacterial rRNAs, but that wheat cytoplasmic rRNA was not. This molecular evidence was powerful support for a bacterial origin of mitochondria and the endosymbiotic theory championed by Lynn Margulis.

A NEW FORM OF LIFE!?

For Woese, these pioneering studies were gratifying validation of the vision he had in 1969, and of the massive effort invested in getting the RNA sequencing effort off the ground. He was just getting started. If rRNA had the power to detect deep relationships and events long in the past, what else could it tell us about the evolution of life?

Woese and his team went on cataloging more species of interest. One of the key limitations they ran into was not deciphering the spots, but growing bacteria that were not common laboratory or clinical species, especially in a low-phosphate medium. As many as half of the species Woese tried were too finicky to grow under their standard conditions, so he reached out to all sorts of experts who had learned how to coax more exotic species to grow under laboratory conditions.

One group he coveted was a bunch of species that produced methane. Based on their different morphologies, some texts placed individual species among different taxa. Woese suspected that the common biochemistry of gas production among the so-called methanogens told a different story. Fortunately for Woese, his Indiana colleague Ralph Wolfe was an expert in the group and just as eager to see what their rRNAs might reveal. Wolfe and his student Bill Balch had worked out a way to grow the methanogens safely in pressurized bottles (methane is explosive!), and George Fox and Balch purified radioactively labeled rRNA from a species called *Methanobacterium thermoautotrophicum*, which everyone called Delta H for short.

On June 11, 1976, Woese began examining the RNA fingerprint as usual—but there was nothing usual about the spot pattern. Right off the bat, he noticed that the two small spots that were in all bacteria were missing. He looked for several other signatures of bacteria, and those, too, were missing. Not all were absent—there were some bacterial oligos, but there were also some eukaryotic oligos as well as some of the universal oligos shared by both bacteria and eukaryotes.

Woese was stumped. What was this RNA? It was not bacterial, nor was it eukaryotic.

Then it dawned on him. Maybe there was something in the world that was neither bacterial nor eukaryotic, perhaps a distant relative of both that no one had recognized before? Woese rushed to tell George Fox, who agreed that maybe there was something else out there . . . something that belonged to a different division than all other species known to biology.

Or perhaps it was just one very strange species. They couldn't just go blurting out such a wild idea without more evidence. So they looked at a second methanogen. The bacterial oligos were missing from that one, too. They took another six months to compile the catalogs of five more methanogen species, all of the various types that had been described. To their great delight, they all told the same story. None of the catalogs were bacterial or eukaryotic. They did, however, all share a unique oligo—a sign that they were closely related to one another.

If these methanogens did not belong to either of the two major divisions of life (bacteria or eukaryotes), and really were a third form of life, Woese and Fox figured that there must be other members of such a major division. Moreover, just as all bacteria and all eukaryotes shared key traits, there should be key traits (beyond their rRNA sequences) that distinguished this third division from the other two. Woese and Fox wondered, what could those traits be?

Woese learned that the cell walls of at least one methanogen were very different from those of bacteria, and that was also true of another group that lived in and required very high salt environments, known as halophiles. Woese obtained a culture of a halophile and examined its rRNA. Bingo. It also lacked the bacterial oligos but shared sequences with the methanogens, which revealed that it belonged to the same, new group.

Bolstered by this additional evidence, Woese and Fox were ready to go public with what they believed to be the profound implications of their discoveries—rather than the two major divisions of life that biologists had accepted for decades, there were three major divisions. They suggested the term *urkingdom* or "primary kingdom" for these three divisions: (i) the "eubacteria," comprised of all typical bacteria; (ii) the eukaryotes, organisms with cell nuclei; and (iii) this new urkingdom, which they dubbed the "archaebacteria." The term denoted their suspicion that the methanogenic group seemed better suited to the environment of the early earth and would be more primitive than the other urkingdoms.

NASA and the National Science Foundation, which had supported Woese's research, issued a joint press release heralding the discovery of the archaebacteria. The media was intrigued, and the story of "a third form of life" made the front page of the *New York Times*.

But the scientific community's response was muted at best. Woese received the same sort of resistance to this new, radical idea that Margulis had a decade earlier. His colleague Wolfe even received a call from one Nobel Prize–winning microbiologist who warned Wolfe to distance himself from Woese's "nonsense" about a third form of life.

Woese had many qualms and reservations to overcome. First, many scientists were unfamiliar with the method of using rRNA to decipher evolutionary relationships. Second, it was hard for some to believe that one molecule could be a reliable record of the past. Some thought it risky to draw such strong conclusions based on one molecule. And third, there was the prevailing, entrenched view that all bacteria belonged to one group (the prokaryotes). Vanquishing those doubts and changing minds would, as for most all revolutionary new ideas, require more evidence, and time.

ARCHAEA AND THE ORIGIN OF EUKARYOTES

Woese pursued more evidence by pursuing more species. He suspected that he might find more archaebacteria among organisms that thrived in extreme environments, like the methanogens and halophiles. He turned to the thermophiles Brock had discovered and described. Sure enough, *Sulfolobus acidocaldarius* and *Thermoplasma* had the telltale rRNA signatures of archaebacteria. The urkingdom was growing.

So, too, was the list of traits that distinguished the archaebacteria. Other researchers discovered a distinct and unusual type of lipid in their membranes, and that the structure of RNA polymerase was very unlike that in bacteria. As the number of archaebacterial species and traits grew, Woese and other researchers were increasingly convinced that they were a third division; they realized, however, that the *bacteria* suffix on the name of the group prompted many scientists to keep lumping them with bacteria. They also wrestled with what to call a classification above the level of kingdom. Woese, Otto Kandler, and Mark Wheelis proposed a name change to "Archaea," and that the group constituted one of three "domains" of life, along with the Eukarya and Bacteria (Figure 10.4A).

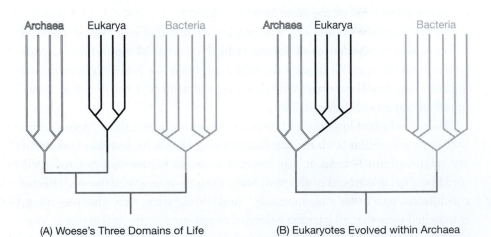

(A) Woese's Three Domains of Life   (B) Eukaryotes Evolved within Archaea

FIGURE 10.4 Woese's Three Domains of Life and the Origin of Eukaryotes (A) On the basis of ribosomal RNA sequence relationships, Woese proposed that the Archaea formed a third domain of life in addition to Bacteria and Eukarya (eukaryotes). (B) Subsequent studies have revealed that eukaryotes arose within the Archaea.

One of the motives behind the name Archaea (and archaebacteria) was Woese's long suspicion that the group was ancient and primitive. If that were true, then deeper study might reveal clues to the evolution of life. Indeed, some of the first studies of archaean molecules turned up some striking resemblances between some archaea and eukaryotes. Proteins archaea use to package their DNA in chromosomes, to transcribe DNA, and to decode RNA messages bore such similarity to eukaryotes that they suggested to some that eukaryotes evolved from some archaeon. Some of these provocative similarities are in short "signature" sequences in proteins that are shared among some archaea and eukaryotes, and no other group. For example, there is a short, 11 amino acid insertion in one protein involved in decoding messenger RNA. This insertion has the following sequence in different eukaryotes and archaea:

Organism	Insert Sequence
Eukaryotes	
Human	GEFEAGISKNG
Yeast	GEFEABISKDG
Tomato	GEFEAGISKDG
Archaea	
Sulfolobus	GEYEAGMSAEG
Pyrodictum	GEFEAGMSAEG
Acidionus	GEFEAGMSEEG
Bacteria	Absent

The existence of this sequence in two domains of life but not in the third would be most logically explained by the archaea and eukaryotes being more closely related to each other on life's tree than to bacteria. The implications for the story of life would be that a common ancestor of all three domains split into two domains, the Bacteria and Archaea, and then the Eukarya arose later from a branch within the Archaea (Figure 10.4B).

Sequencing of eukaryotic genomes, however, suggested that many eukaryotic genes were more related to those of Bacteria than of Archaea. This raised the strong possibility that the first eukaryotes were the product of a mixed marriage—a fusion of archaeal and bacterial parents. The once tenuous endosymbiotic origins of eukaryotes now have a very specific pedigree. Further molecular studies have identified the likely specific archaeal parental host cell among a branch called the Lokiarchaeota, and the specific bacterial symbiont among the alphaproteobacteria.

Woese's hunch that the Archaea are both ancient and integral to the story of life has been proved beyond any doubt. His pioneering molecular studies not only revolutionized the study of evolutionary relationships but, as his friend and fellow archaeal enthusiast Otto Kandler said, "He opened a door which nobody expected to exist."

So, if you happen to visit magnificent Yellowstone, don't turn away from the stinking, boiling soup of those hot cauldrons or be revolted by the strings and mats of slime that ooze around their edges. That's no way to respect one's relatives, no matter how distant. Ponder the amazing fact that you share hundreds of genes with members of this community. And, in this sort of community, somewhere an unfathomably long time ago, perhaps along a deep sea vent, there emerged in a belch of methane the ancestor of all of the familiar and visible kingdoms on earth.

END-OF-CHAPTER QUESTIONS

1. Why was bacterial genealogy in such a poor state prior to Woese's work?
2. Why was ribosomal RNA a good record of evolutionary genealogy?
3. What initial evidence led Woese and Fox to consider the possibility that methanogens were a third form of life distinct from bacteria and eukaryotes?
4. Explain the significance of archaea and eukaryotes sharing genes and sequences that are not shared with bacteria.
5. Draw parallels between the scientific resistance to and ultimate acceptance of Woese's third kingdom of life and the endosymbiotic origin of eukaryotic organelles. What factor or factors tipped the balance to acceptance?

11

QUEEN OF THE STONE AGE

Man is a tool-making animal.
—BENJAMIN FRANKLIN

THE PASSAGE INTO THE cave was long and very narrow. Crawling on her hands and knees, 12-year-old Mary Nicol was just slim enough to squeeze through the low and tight spaces, her path lit by the lamps she, her mother Cecilia, and their guide Abbé Lemozi carried. Pêch Merle Cave had been discovered just a few years earlier in 1922 by a couple of teenage boys, near the Abbé's village of Cabrerets, one of a constellation of prehistoric rock shelters and caves in southwest France that the Abbé had explored and studied.

Getting into the cave was so difficult, it was not yet open to the public. Cecilia gashed her head on a wall and started bleeding, but the excitement of glimpsing what few people, or at least modern people, had ever seen propelled the trio on. After a long crawl, the passageway opened up into the cave itself. In the lamplight, their perseverance was rewarded. The walls and boulders were covered with magnificent paintings of bison, mammoths, and spotted horses. Here and there were handprints of the artists themselves, simple stencils in silhouette. Mary was enthralled. She wondered about the age of the artworks, and the world of their makers (Figure 11.1).

It was Mary's father, Erskine, who introduced Mary to the treasures within France's famous caves. Outside those near Les Eyzies, he and Mary had permission to sift through the heaps of debris excavated by archeologists. In the

FIGURE 11.1 Pêch Merle Cave Drawing The spotted horses enchanted young Mary Nicol. Prisma Archivo/Alamy Stock Photo.

discarded rubble, Mary found ancient tools with chiseled points and end-scrapers used for cleaning animal hides.

Mary's father, Erskine, was a talented, professional landscape painter. His work determined the course of the family's bohemian lifestyle. They traveled together through Italy, France, or Switzerland for most of the year, and went to London to sell the results. The family's constant travel meant no regular schooling for Mary, which suited her just fine. She learned Italian and French, and she took advantage of every opportunity to visit caves and become the family's chief collector of stone artifacts.

But when she was just 13, those opportunities, and the family's happy wandering ways, came to an abrupt halt. Her father, whom Mary adored and believed to be "the best person in the world," suddenly fell ill and died. Mary was crushed and inconsolable.

WHAT TO DO WITH MARY?

Money was usually tight for the Nicols. Now faced with the loss of the family breadwinner, Mary's mother decided that they had to return to England. The question arose of what do about Mary's education.

The answer was a convent school in London, or rather a series of convent schools because Mary, raised as she was, did not take well to the confines of school or particular subjects. She was dismissed from the first for hiding frequently in a boiler room, and refusing to read poetry aloud to an assembly. At the second, she faked a fit in a classroom, even using a bar of soap to help her froth at the mouth. That prank was not sufficient for expulsion, but setting off a loud explosion in a chemistry class was. Mary would later say, "At least I ended my school career with a bit of a bang."

Her mother was forced to give up any hope of a regular education; however, Mary's interest in archeology and art remained. When she was 17, she attended lectures on archeology and took some courses at University College. Since she had never passed a single school exam, Mary had no chance of getting into a university or earning a degree. So the challenge was to find some kind of career training. For archeology, that meant getting some kind of real field experience.

Mary wrote to anyone she heard about who was conducting excavations. Most refused politely, but Dorothy Liddell offered her a spot as an assistant working on a dig in southern England. Liddell appreciated Mary's work ethic, and asked her to draw some of the finds for publication. The experience was an eye-opener for Mary, not just in terms of learning the necessary field skills, but in seeing that, in 1930, a woman could be in charge of an excavation and be taken seriously as an archeologist. And it was not just one woman, for Liddell introduced Mary to Gertrude Caton-Thompson, a leading British scholar on Egypt who had unearthed stone tools in the desert west of the Nile River. Impressed by what she saw of Mary's artwork, she asked Mary to draw illustrations for a book on her Egyptian work.

In turn, Caton-Thompson wanted to introduce Mary to an up-and-coming archeologist who was looking for an artist to draw some of his discoveries. Louis Leakey had been making headlines with expeditions he had led in East Africa. Gertrude brought Mary as her guest to one of Leakey's frequent lectures around London, and arranged for her to sit next to him at the dinner that followed. The encounter would soon change their lives and, eventually, the story of our species.

THE QUEST FOR THE TOOLMAKER

Louis Leakey was born in Kenya to British missionaries—he considered himself more Kikuyu than British. Although the first white baby that most had ever seen, Louis was welcomed into Kikuyu society. As a boy, he learned the ways of the Kikuyu firsthand. He was accepted as a brother by the tribe, taught how to throw a spear, to wield a war club, and to track, hunt, and trap wild game.

But few experiences stoked Louis' interests more, or had a greater impact on his future direction, than a simple children's book he received as a Christmas present from a relative in England. Entitled *Days before History*, the book recounted the tale of a young boy named Tig in the Stone Age of Britain. The drawings of the flint arrowheads and axes of this culture captivated Louis. Convinced that Stone Age people must have also lived in his region of Kenya, he began collecting pieces of rock with any resemblance to the flint tools he read about. His family teased him about what they called his "broken bottles."

One day, Louis timidly showed his collection to a family friend who was curator of Nairobi's Natural History Museum. The curator confirmed that some of his stones were indeed ancient tools. Louis was thrilled as the man explained that Kenyan Stone Age cultures made tools out of obsidian, a glass-like volcanic rock, because the flint used in tools elsewhere was not available in East Africa. Louis became obsessed with collecting stone tools. He learned how little was actually known about the Stone Age, particularly in East Africa. At age 13, he made up his mind that this would be his life's quest.

Seventeen years later, Louis was convinced that those tools were clues to one of the greatest mysteries in all of science—where humans originated. He believed, as Darwin had first reasoned based on our greater similarity to gorillas and chimpanzees than other apes, that humans had likely evolved in Africa from ancestors we shared with gorillas and chimps.

In the early 1930s, however, most scientists thought that Asia was the birthplace of our species, largely because of hominid fossils that had been found in Indonesia and China. Nevertheless, Leakey believed that the countless primitive tools he had found in East Africa were evidence that humans had an earlier history in Africa. And he was determined to prove it by finding the bones of the ancestors who made those tools. His forthcoming book *Adam's Ancestors* was to help make that case.

At the dinner, Mary agreed to help illustrate the book. Over the ensuing months, the two saw a lot of one another. Although Louis was 10 years older, and married with one child and another on the way, Mary was attracted by Louis' boundless self-confidence, his love for wild places, working in the field, and being near wild animals. And although Mary was just 20, her unconventional upbringing, rebellious streak, artistic talent, and passion for prehistory were irresistible to Louis. To the dismay of Mary's mother, her mentors, and Louis' parents, Mary and Louis fell in love, and Louis proposed. Mary accepted, and arranged to join Louis on his next expedition, where she also fell in love with Africa.

OLDUVAI

In 1935, just getting to the African bush was half the adventure.

Mary arrived by plane in Moshi, Tanzania, where Louis met her to make the long trek to one of the most promising fossil sites he had scouted, Olduvai Gorge, which was on the edge of the Serengeti Plain. To get there, however, the route went up and over the forested slope of Ngorongoro Crater, a giant extinct caldera. The road, if one could call it that, had only recently been made, and ascended 2000 feet to the crater's edge in a series of switchbacks. The climb up the muddy, 16-mile-long road took two and one half days.

After resting and getting cleaned up, the couple then descended toward Olduvai and the Serengeti. Mary's first glimpses etched indelible images of what was to become her home:

> Starting the descent from Ngorongoro to the Serengeti . . . I was looking spellbound for the first time at a view that has since come to mean more to me than any other in the world. As one comes over the shoulder of the volcanic highlands to start the descent, so suddenly one sees the Serengeti, the plains stretching away to the horizon like the sea, a green vastness in the rains, golden at other times of the year, fading to blue and gray. . . . In the foreground is a broken, rugged country of volcanic rocks and flat-topped acacias, falling steeply to the plains. Out on the plains can be seen small hills . . . the scale is so vast that one cannot tell that the biggest is several hundred feet high. Olduvai Gorge can also be seen. Two narrow converging dark lines, softened by distance and heat haze, pick out the Main Gorge and Side Gorge. . . . I shall never tire of that view, whether in the rains of the dry season, in the heat of the day or in the evening when one is driving down straight toward the sunset. It is always the same; and always different.

For three months, Louis and Mary walked and sometimes crawled through Olduvai Gorge, scouring every inch they could for tools, a hint of a toolmaker,

FIGURE 11.2 Olduvai Gorge This gorge contains a rich record of the last 2 million years of human history. Photo courtesy of Patrick Carroll.

or other fossils. Every exposed hillside or gully yielded some sort of archeological or geological treasure, including a delicate pig skull, a herd of gazelle-like skeletons, and the massive bones of an elephant. Tipped off by a local Masai that there were more "bones like stone" nearby at a place called Laetolil (now called Laetoli), reconnaissance confirmed abundant fossils there. There were also quite live inhabitants as well. One morning, though Louis had warned her not to stray off on her own, Mary nearly stepped right on top of a sleeping lioness. She later explained, "She and I were mutually horrified and fled in opposite directions. . . . Meeting a lioness on foot is not as disastrous as it may sound, unless she has her cubs with her" (Figure 11.2).

Getting and maintaining supplies at such a remote place was quite a challenge. They had to rely on the gorge for water, which was often scarce. Mary found that it was fairly drinkable after a fresh rain, if a bit silty. Once it had been standing for a while, however, and used by the wildlife "it becomes the most unpleasant, since many animals, including rhinos, often urinate at waterholes. An attempt to filter this water through charcoal met with no success, so that our soup, tea, or coffee all tasted of rhino urine, which we never got used to."

Although they found no hominid bones, tools were abundant, including many impressive hand axes. Louis was confident, writing in his monthly field report, "I still am convinced—that somewhere at Olduvai we shall sooner or later find the fossilized remains of the men who made the . . . tools which are so plentiful."

After they left the gorge, Louis took Mary to Kondoa, where there were many paintings on various rock shelters. Mary was entranced by the beautiful human and animal figures and made many tracings of them. The rock art, the dramatic landscape, the beautiful wildlife, the fascinating people—the whole continent had "cast a spell on her."

After an exhilarating journey, Louis and Mary returned to England, where they married and plotted their next adventure in Africa.

A STONE AGE FACTORY

When the Leakeys returned to Kenya, they were nearly broke. In order to continue working in the field and living in East Africa, some means of support was necessary. Louis, already a prolific writer, committed to writing a history of the Kikuyu tribe. Mary just wanted to excavate. Anything would do—she was not concerned about the age or place; she was interested in everything. She hurled herself into several sites around the Nakuru region.

The partners' roles shifted. Louis wrote more, lectured, and took up administrative duties with a Kenyan museum. It helped to pay their expenses and to create a space for their collections. Mary loved the field, and relished every day spent excavating—living in a grass hut in the bush satisfied her sense of adventure. Mary quickly proved herself an extraordinary archeologist, much more methodical and meticulous than Louis. Amazingly, from one trench alone she recovered and cataloged more than 75,000 tools from the late Stone Age.

For security, she obtained a Dalmatian and so loved the breed that from that day forward she would never again be without one or more dogs. She also developed a love for whisky and Cuban cigars as small rewards at the end of a long workday.

Then, children arrived. Their first, Jonathan, was born in 1940. Louis and Mary could manage only short forays into the field. On Easter weekend 1942, they journeyed to Olorgesailie, about 30 miles southwest of Nairobi. Two decades earlier some tools had been reported from the area, but the details

FIGURE 11.3 Tool Factory at Olorgesailie Mary and Louis Leakey discovered a field of hand axes and choppers at Olorgesailie. Photo courtesy of the Human Origins Program, Smithsonian Institution.

and location were sketchy. Louis and Mary, accompanied by a few staff, spread out over the white sediments. At almost the exact same moment, they called out to one another. Mary kept shouting at Louis to hurry over to see what she had found. He marked his place, and went to Mary. "When I saw her site I could scarcely believe my eyes. In an area of fifty by sixty feet there were literally hundreds upon hundreds of perfect, very large hand axes and cleavers" (Figure 11.3). Mary thought the scene looked as though the place had only just been abandoned by the toolmakers. Louis guessed the "factory" site was 125,000 years old, a figure that was scoffed at as too great (subsequent radiometric dating put the age of the site at more than 700,000 years old).

The scene was so startling and impressive that they decided to leave a large section exactly as they found it. A catwalk was built, the site was opened as a museum in 1947, and Olorgesailie remains a national museum today.

Mary's subsequent careful and exhaustive excavations of individual layers revealed tools, animal bones, and in some cases arrangements of stones where shelters probably once stood. Olorgesailie was chock-full of animal fossils and tools but, alas, no remains of the toolmakers turned up.

APE ISLAND

By the late 1940s, the Leakeys' search for human ancestors in East Africa was now more than 20 years running. While the tool record was fabulous, they had little human fossil material to show for their efforts. They had been searching in progressively older sediments. There was another approach to consider: start further back in time and try to work forward. In other words, look for fossil apes that might illuminate the split between the ape and human branches of the primate tree.

There was widespread bias that humans were so different from the great apes that the split had to be pretty deep in time in the Miocene, the geological epoch that opened about 23 million years ago. Louis had previously found a lot of Miocene fossils in short visits to Rusinga Island in Lake Victoria, and he imagined he could do much better if the place were properly excavated.

The whole family, including their two children, Jonathan (8) and Richard (4), went along. Rusinga became a favorite spot for the Leakeys' Christmas holiday. Well, it was a holiday for the kids at least. Mary laid down bedding for the boys on top of the supply crates in the back of the family Dodge, and they would set out before sunrise for the Rift Valley and beyond. It was an all-day, 400-mile trip to Kisumu on the shore of Lake Victoria, then an overnight boat ride to the island.

The family lived on the boat and explored the island by day. The boys entertained themselves by fishing, playing with the local children, and occasionally finding fossils. At age 6 Richard found his first: the complete jaw of an extinct pig. Unlike many other sites, there was fresh water to bathe in, if one did not mind the crocodiles. The family routine was for everyone to get ready to jump in, then Louis would fire a couple of shotgun blasts into the water, which he believed ensured 15 minutes of safety.

In the afternoons, Louis often took the boys on long walks to go prospecting for fossils. He pointed out birds and butterflies, showed them how to sneak up on wild game, to make tools out of stones, and to make fire by rubbing sticks together. The boys cherished these adventures more than anything else and each became keen naturalists (Figure 11.4).

Mary, for her part, "never cared in the least for crocodiles, living or fossil." She went looking for something more interesting. One day, she saw some promising bone fragments lying exposed, with a tooth protruding from a slope. "Could it be . . . ?" she wondered. She called to Louis and he came running. As they brushed away the loose sediments from the tooth, a jaw emerged. Even better, it was clear that a good portion of the creature's face was present.

FIGURE 11.4 The Leakey Clan A family portrait of (left to right) Richard, Mary, Philip, Louis, Jonathan, and the pack of family Dalmatians. Leakey Family Archives.

This was the first skull of a fossil ape of any age to be discovered. A great prize, but first it had to be put together. Mary worked long hours to piece together the more than 30 separate pieces of the skull, some as small as a match head. She and Louis were the first to see what the ape, called *Proconsul*, looked like.

As Louis spread the news of their find, their British colleagues were eager to see it for themselves. Louis thought Mary, as finder of the skull, should take it to England.

Proconsul and Mary both got VIP treatment. The airline offered her a free flight and *Proconsul* traveled in a box on her lap. When Mary arrived at London's Heathrow Airport, the press was there en masse. They asked her to pose on the gangway. She and the fossil ape were, much to Mary's surprise, front-page news. Unaccustomed to any attention, Mary was relieved to hand the skull over for further study at Oxford, and to return to Africa.

While Mary had done her best to dodge the limelight, the *Proconsul* publicity had a very important tangible dividend beyond its scientific value—it brought attention and funding to the Leakeys' efforts that would enable them to sustain excavations, not just at Rusinga, but finally at Olduvai.

DEAR BOY

Louis and Mary had scouted Olduvai, but they had not undertaken a proper excavation. The reasons for this were twofold. First, the gorge was very large. They had explored 180 miles of exposures that ranged from 50 to 300 feet in depth. It was important to be selective about where to dig, and there were a very large number of promising sites. The second was the shortage of funds, and the necessity of doing other things to make ends meet.

The success of *Proconsul* attracted grants and benefactors. One of the latter was Charles Boise, a London businessman with a keen interest in prehistory. He had funded part of the Rusinga expedition and now committed to supporting the Leakeys for seven years. Thanks to Boise's generosity, they decided to excavate at Olduvai, determined to find an early human.

The excavations went on almost every year throughout the 1950s and focused mostly on what was called Bed II, a lower site that yielded more than 11,000 artifacts and enormous numbers of large mammal fossils. The fossils were exceptionally well preserved. These included, for example, complete skulls of a giant buffalo-like animal called *Pelorovis* with horns spanning more than 6 feet. Moreover, they found a pile of *Pelorovis* bones that had been butchered by the tool-wielding inhabitants of that time. Yet, the hominids themselves were ghosts, as only two teeth were found in a seven-year span.

In 1959, they turned their attention to Bed I, the oldest of the Olduvai beds. On the morning of July 17, Mary went out prospecting alone while Louis remained back in camp sick. Mary saw a lot of material lying out at the surface, but then one scrap of bone that was projecting from beneath the surface caught her eye. It looked like part of a skull—a hominid skull. She carefully brushed away a little of the sediment and saw parts of two large teeth in the upper jaw. They *were* hominid for sure, she thought. Mary had a hominid skull, and a lot of it, at that. She jumped into her Land Rover and raced back to camp shouting out, "I've got him! I've got him! I've got him!"

Louis asked, "Got what? Are you hurt?"

Mary blurted out, "Him, the man! Our man. The one we've been looking for. Come quick."

Louis recovered instantly from his illness and raced back with Mary to the site. Louis saw that it was a hominid, and that most of the skull was there. After 24 years of searching East Africa, they had indeed found "their Man."

Their emotions would soon be shared with and by their many colleagues in paleoanthropology.

FIGURE 11.5 Dear Boy *Zinjanthropus*, now known as *Australopithecus* or *Paranthropus boisei*. John Reader/Science Source.

First, they had to extract and piece together what Mary came to call her "Dear Boy." The skull was in some 400 pieces. Mary, as she had with *Proconsul*, patiently put him together. She had the upper jaw and teeth, most of the face, and the top and back of the skull. He was, in Louis' words, "quite lovely" (Figure 11.5).

Deciphering where this fossil fit in the scheme of human evolution was a huge challenge, as there was relatively little to compare it to. It bore some resemblance to the so-called australopithecines that had been found previously in quarries and caves in South Africa. Though none of the australopithecine skulls were as complete, Dear Boy's similarities posed a conundrum because Louis had thought of them as an evolutionary dead end, an offshoot of the line that gave rise to modern humans. Furthermore, no evidence of australopithecine toolmaking had been unearthed. Now, he was looking at a skull that had come from tool-bearing deposits. The brain was too small to classify Dear Boy as a member of our genus *Homo*. A new name also certainly would help make their discovery stand apart from others. He chose the name *Zinjanthropus* ("man from East Africa") *boisei* (after their benefactor).

Louis dashed off a note to the journal *Nature* describing the new hominid and could not wait to spread the news. Newspapers around the world blared headlines of the great discovery. Louis embarked on a speaking tour that included a triumphant night in London and 66 talks at 17 universities across the United States. Audiences were captivated by Louis' account of the Leakeys' long search for and ultimate success in finding Zinj.

The question of Zinj's age was on everybody's mind. How far back did our hominid ancestry reach? Louis told his audiences that he thought Zinj lived more than 600,000 years ago. This figure was based on study of the Olduvai sediments and estimates about their rate of formation. But shortly after Zinj was discovered, two geophysicists used a new potassium–argon dating technique to obtain a figure of 1.75 ± 0.25 *million* years for the ash bed just above where Zinj was found. This date was staggering—Zinj was three times older than Louis had thought, which some had believed to have been an exaggeration. The tools and the toolmakers were in fact far older than anyone had imagined.

The discovery and dating of Dear Boy changed the course of paleoanthropology. They finally brought everyone's undivided attention to Africa as the birthplace of humanity, where Louis (and Darwin) thought it belonged. And they brought the time frame of human evolution into real terms.

They also changed the course of the Leakeys' lives and work.

The National Geographic Society gave the Leakeys the largest grant they had ever received and the American public was so enthralled that they, too, began to support the Leakeys' research. The long drought for hominids, and for funding, was over and a whole new era of paleoanthropology was opening.

THREE SPECIES!

The next year at Olduvai (1960) the excavation opened with a great sense of anticipation. Louis, now caught up in so much public relations and museum work, could only visit from time to time. Mary, accompanied by her Dalmatians, set up a permanent camp and led the team. It was a massive undertaking: the excavation was on an unprecedented scale that involved more hours of labor in 1960 alone than in all of the nearly 30 previous years put together.

The children were now much older. Jonathan, 20, worked at Olduvai for a few months. One day he asked his mother, "Does any animal have a long thin bone like this?" tracing a shape through the air with his finger. Mary replied she couldn't think of one and Jonathan casually said, "Oh, then I think it must be a hominid." Mary dropped her work and rushed to see what Jonathan was talking about. Sure enough, it was a hominid leg bone—a fibula. Later he found a tooth and toe bone. Mary decided to excavate "Jonny's site."

There, Jonathan found the remains of two individuals, including a skull. While just 100 yards from Zinj, the skull was about a foot below the level where Zinj was found, so it was older. And it was different. The skull enclosed a larger brain, and its shape was more similar to a modern human. Amazingly, the excavation turned up 21 bones of a hand, and 12 bones of a foot. It was no doubt a different species from Zinj. Louis hoped this new find might be closer to our *Homo* line than Zinj. The tips of the fingers and thumbs indicated this hand was quite capable of making the tools found nearby (Figure 11.6). This new hominid was closer to our species and was dubbed *Homo habilis*, meaning "handy, able, or skillful man." This meant there were two lines of hominids at Olduvai. The human tree was growing, and branching out.

Louis was not completely left out of the fossil-finding at Olduvai. Late in the 1960 season he, 11-year-old Philip, and a geologist went out prospecting in the gorge. Louis spotted what he first thought was a tortoise shell. He was happily wrong. It was a hominid skull. He fetched Mary and beamed over his prize. Excavation revealed this skull was different from either Dear Boy or *Homo habilis*, but bore a strong resemblance to *Homo erectus* from Asia. Dating revealed it to be much older than the Asian fossils, at 1.4 million years old. The three fossils meant that *three* different species of hominids had lived at Olduvai over the span of just a few hundred thousand years.

With a vast collection of Stone Age tools and the discovery of the oldest known hominids, Louis and Mary reigned atop the world of paleoanthropology until Louis' death in October 1972. But even then, Mary's adventures were far from over.

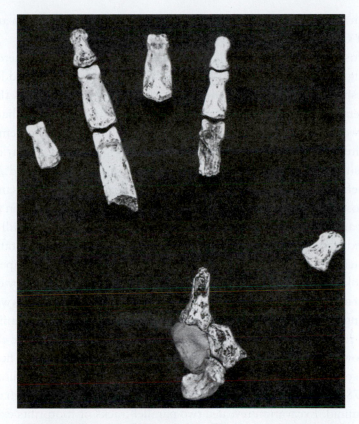

FIGURE 11.6 Hand of *Homo habilis* The bones reveal this hand had the precision grip necessary for fine manipulation of objects.

Day (1976). Courtesy of Professor Michael Day.

SOMETHING FOR THE MANTELPIECE

Mary's years of excavations at Olduvai yielded tens of thousands of artifacts, including the oldest known tools and a very diverse array of implements—choppers, chisels, cleavers, scrapers, picks, and more—of varying sizes and shapes. As early as 2 million years ago, hominids were fashioning specific tools for specific purposes. The diversity of hominids and tools found at Olduvai led to the inescapable deduction that there must have been yet older cultures in East Africa. Her excavations had, however, reached the bottom of the Olduvai fossil beds. Clues to those older cultures had to be sought elsewhere.

Laetoli was about 30 miles from Olduvai. Over the Christmas holidays in 1974, Mary and her team found a hominid jaw and some teeth there. Those fossils turned out to be 3.3 to 3.7 million years old, much older than anything found at Olduvai. Mary moved her operation to Laetoli.

The fossil-hunting was good, but the working conditions at Laetoli were even tougher than at Olduvai. The area was chock-full of puff adders, the snake responsible for the most deaths in all of Africa. Sometimes the team had to remove two snakes a day from camp. The buffaloes and elephants also did not welcome the human instrusion, and chased several team members.

Despite these drawbacks, the camp drew a lot of visitors. One day in 1976, three visiting scientists—Jonah Western, Kaye Behrensmayer, and Andrew Hill—went for a walk in a dry riverbed. Elephants had used the same path and deposited many cannonball-sized droppings that had dried out in the sun. Western and Hill got into a playful elephant dung–throwing fight. Hill ducked down and when he looked at the ground, he saw a pattern of indentations on the surface that reminded him of raindrops preserved in ash at Pompeii, Italy. As he looked closer, he saw crisp impressions of animal tracks. Mary directed the team away from searching for bones to searching for more trackways and excavating them. The tracks were all over and of many kinds of animals— giraffes, rabbits, antelope, elephants. Apparently, a nearby volcanic eruption was followed by rainfall that cemented the fresh prints in place. These were quickly covered by more ash and preserved undisturbed for 3.5 million years.

In 1978, however, some prints were uncovered that were not animal—they were hominid. Excavation revealed two parallel tracks of footprints extending for about 80 feet. One set of footprints was smaller than the other, indicating that a juvenile or female was alongside an older or male individual. Skeletons, leg bones, and foot bones, which can indicate whether an individual walked upright or not, were and remain very hard to come by for ancient hominids— and none had been found at Laetoli. Here, Mary stared at the most vivid evidence imaginable of the bipedal stride of our ancestors 3.5 million years ago (Figure 11.7).

Africa had saved her best for last. After excavating a particularly sharp set of prints herself, Mary lit a cigar and, admiring the impressions, declared, "Now this is really something to put on the mantelpiece."

Although Mary never attended college or received any formal training, her pioneering achievements in archeology and her discoveries of several of the most important homind fossils ever found earned her honorary doctorates from universities on three continents.

FIGURE 11.7 Something for the Mantelpiece Mary Leakey at the end of the trail of hominid footprints discovered at Laetoli.

John Reader/Science Source.

END-OF-CHAPTER QUESTIONS

1. Why was Louis Leakey convinced that Africa was the cradle of humanity?
2. Why might stone tools be easier to find than hominid bones?
3. Why were the Laetoli footprints such important evidence about early hominid locomotion?
4. Despite having found Zinj and reassembled the fossil from several hundred pieces, Mary Leakey was not a coauthor of the paper reporting the discovery. The one time she is referred to is at the beginning of the report: "On July 17, at Olduvai Gorge in Tanganyika Territory, at site FLK, my wife found a fossil hominid skull. . . ." Why do you think Mary was left out of the authorship of the discovery?
5. What characteristics did Mary Leakey have that contributed to her success as a scientist?

12

A LITTLE BIT OF NEANDERTHAL IN MANY OF US

*Genes are like the story, and DNA is the language
that the story is written in.*

—SAM KEAN

THE SCENE WAS PART Cold War thriller, part science fiction horror movie.

It was the summer of 1983. Berlin was still a divided city. East Berlin, the capital of the German Democratic Republic (GDR), was separated from West Berlin and a free Europe by a 26-mile-long wall (the wall did not come down until November 1989). The communist GDR was under the firm grip of the Stasi, the secret police, and its borders tightly controlled. Anyone who sought to escape the East did so at the risk of being shot, and several had been killed in recent years.

Some Westerners, however, could get passes to visit the GDR. Svante Pääbo, a 28-year-old Swedish graduate student, was on a mission in the heart of East Berlin. His destination was the Bode Museum, one of several in a complex on an island in the middle of the River Spree. It still bore the scars of the Soviet Army's assault during the battle for the city at the climax of World War II, 40 years earlier. Bullet holes pocked the walls, and the roof of the exhibit hall had huge openings from artillery shells.

After passing through several security checks, Pääbo was ushered into the storage facility where the curator showed him what he had come for—a row of dead bodies. But these were no ordinary corpses that Pääbo had gone to so much trouble to see; they were several thousand years old. They were Egyptian mummies.

None of them were in the condition one would see on display; instead, these mummies were all unwrapped and broken. Not good enough for museum exhibits, they were just fine for Pääbo, who wanted to collect little snippets of tissue to test a wild idea: Could he get DNA from 2000- to 5000-year-old mummies? And if he could, what could that tell him about these ancient people that archeologists and Egyptologists did not know?

Pääbo had wisely kept his visit to East Berlin secret, lest his PhD supervisor back in Sweden think he was wasting time and resources, and cut him off.

FROM MUMMIES TO MOLECULAR BIOLOGY, AND BACK AGAIN

Ever since his mother took him to Egypt when he was 13, Pääbo had been enthralled with ancient history. He enrolled at Uppsala University determined to become an Egyptologist, but soon discovered that academic life was not going to be all about pyramids and pharaohs. He decided to switch to medicine, and then to pursue a PhD studying the molecular genetics of the immune system. But his fascination for all things Egyptian never faded. He continued to go to lectures, take classes in Coptic (the language of Egypt during the Christian era), and meet and befriend experts. As he started to learn the new tools for cloning DNA, he wondered whether techniques for DNA forensics could also be applied to ancient humans.

Nobody knew. Pääbo scoured the scientific literature and could not find any reports of the isolation of DNA from anything ancient, human or otherwise. So it was up to him to give it a shot. He first tried to simulate the conditions of mummy tissue by desiccating a piece of calf liver in a lab oven. The horrid smell soon gave away his experiment, although he was able to recover DNA. A friend then gave him access to a few mummies in Uppsala. Mummy tissue turned out, however, to be a lot tougher than dried calf liver. He could not find intact DNA in any tissue samples. He hoped that might just be a matter of preservation. Perhaps he needed to sample many mummies to find one with better tissues and DNA.

Once Pääbo learned that the Bode Museum had a huge collection, the right introductions were made, and that's how he found himself going back to the museum day after day for two weeks, collecting samples of all sorts of body tissues from 23 different mummies.

Back in the lab at Uppsala, Pääbo examined his specimens at night and weekends to avoid any scrutiny from labmates. As a first test of whether any tissues might contain DNA, he rehydrated them and then stained thin slices of tissue with ethidium bromide, a dye that binds to DNA and fluoresces under ultraviolet light. Disappointingly, most samples failed to stain.

One night, he examined a thin section of ear cartilage. Just as in bone tissue, cells in cartilage are surrounded by hard tissue. When he looked in the microscope, he was thrilled to see cell nuclei lit up.

Just three mummies in all stained positive for DNA, but that was enough to try to isolate it. The skin from the left leg of a child showed the most stained cells, so he extracted DNA from the tissue. He then cloned and determined the partial sequence of a ~3400 base pair segment of DNA. The sequence was definitely human; it contained a segment that was frequently repeated in human DNA. (He later realized the segment could be a contaminant from people who handled the mummy, but this possibility has not been confirmed or excluded.) Carbon dating revealed that the mummy was about 2400 years old.

Pääbo's achievement was a first, but before he could tell the world, he had to tell his thesis advisor. With some anxiety, he told his professor what he had done, and showed him a manuscript that he had written. To Pääbo's great relief, his advisor was not at all upset, but was very supportive. In the very prestigious journal *Nature*, Pääbo suggested that his results "raise the hope that recombinant DNA techniques may be applied systematically to archeological samples."

Once he secured his PhD, Pääbo abandoned medical research and returned to his first love of ancient history.

OF MAMMOTHS, MOAS, AND HUMANS

Just as Pääbo committed to the new field of ancient DNA, a huge technical breakthrough promised to bring the past into his and others' reach. Hunting for ancient DNA was hard, but getting information out of that DNA was even harder. It required the cloning of longer stretches of intact DNA so that they could be sequenced. The mummies revealed that ancient DNA was perishable. So when a new technique called the polymerase chain reaction (PCR) appeared that could quickly and specifically amplify segments of DNA from tissues a millionfold or more, Pääbo was eager to apply the method to all sorts of museum specimens.

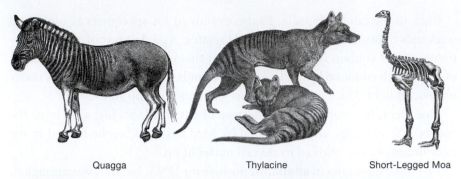

Quagga   Thylacine   Short-Legged Moa

FIGURE 12.1 Extinct Sources of Ancient DNA: Quagga, Thylacine, and Moa

Lydekker (1904). World History Archive/Alamy Stock Photo; Antiqua Print Gallery/Alamy Stock Photo; Ivan-96/Getty Images.

With human specimens scarce, he thought it was wiser to work out the technique on readily available and less precious specimens. The quagga was a South African member of the zebra family that went extinct in 1883 because of overhunting (Figure 12.1). The only quaggas left in the world are stuffed museum animals. Pääbo figured that because mitochondrial DNA (mtDNA) is present in thousands of copies per cell, he would have a better chance of recovering pieces of it. He extracted DNA from the skin of one specimen and used the new PCR technique to amplify pieces of mitochondrial DNA. The method worked beautifully, so well that Pääbo was able to repeat the experiment again and again.

A whole zoo of extinct creatures was opened up to molecular study. Pääbo teamed up with a succession of collaborators to analyze DNA from the thylacine, a marsupial carnivore that inhabited Australia until the 1930s; from the moa, an extinct flightless bird from New Zealand; from ice age horses; and from Siberian mammoths that were up to 50,000 years old (see Figure 12.1). The studies provided two important dividends. First, the DNA sequences from the extinct species could be compared with living relatives and became an entirely new means of sorting out evolutionary relationships between species. Biologists were no longer limited to inferences drawn from the anatomy of extinct species. For example, Pääbo and his collaborators were able to establish that moas were not closely related to kiwis and therefore that flightlessness evolved more than once in New Zealand.

The second important contribution of ancient animal DNA was technical. As more and different kinds of specimens were studied, Pääbo and colleagues gained a better understanding of the fate of DNA in dead creatures. It turned out that DNA was usually not preserved over thousands of years, and when it was, it was generally degraded into pieces that were 100–200 nucleotides long, much smaller than individual genes. Chemical modifications also occurred to individual bases post mortem that introduced variations in sequences. And more challenging still, most of the DNA in specimens was not from the species of interest—it was from bacteria and other organisms that invaded the tissue after death. Nevertheless, with improvements in techniques for extracting DNA, and refinements in PCR for amplifying animal sequences specifically, Pääbo and his team were able to peer back in time to the animals of the last ice age.

As exotic and fascinating as these beasts were, they were not Pääbo's main quarry. His goal was to study human history. Pääbo and colleagues obtained samples of 7000-year-old humans from Florida sinkholes, and from two 5000-year-old bodies frozen in the Alps. They were able to extract and amplify DNA. But it became clear that with respect to human samples, PCR was both a blessing and a curse. PCR could amplify tiny amounts of human DNA, but in ancient samples, that tiny amount could be a remnant either from the specimen itself or from modern humans who handled the sample.

Pääbo was alerted to the contamination issue by discovering that human DNA could also be amplified from many museum animals. The root of the problem was underscored when Pääbo was seeking to obtain DNA from *Mylodon* bones collected by Darwin (see Figure 7.2). He asked the curator whether the bones had been varnished. To Pääbo's horror, the curator picked up the bone, licked it, and said, "No, these have not been treated."

The power of PCR that enabled forensic scientists to detect minute quantities of human blood, skin cells, or saliva at a crime scene was a bane for archeological specimens that may have been handled by untold numbers of people over the decades.

How could he tell that a DNA sequence came from the specimen and not a contaminant? It was very hard to tell the difference. Pääbo calculated that new mutations accumulated so slowly that perhaps one mutation at most would have occurred in 250 generations, or 5000 years.

Pääbo was very frustrated. It was straightforward to distinguish animal sequences from other species in an animal specimen, and so that work flourished. But human studies were fraught with difficulties. A decade after his initial success with mummies, he decided to abandon all work on humans.

His resolve did not last.

It soon dawned on Pääbo what he needed to do—go after humans who were so old that their DNA sequences had to differ from modern humans. Indeed, he needed a different species altogether.

NEANDERTHAL MAN

Eight miles east of Düsseldorf, Germany, the Düssel River cuts through deposits of limestone laid down over 300 million years ago. Named in honor of Joachim Neander, a 17th-century poet, teacher, and composer of hymns, the Neander Valley was dotted with caves and rock shelters until the mid-1800s. Demand for limestone led to quarrying operations on a large scale during which entire caves on the valley walls were removed.

One day in August 1856, the clay floor was being cleared from a small grotto called Feldhofer Cave to prevent its contamination of the high-grade limestone. As chunks of clay were tossed some 60 feet down to the valley floor, a number of bones and part of a skull were exposed. One of the owners of the quarry told the workers to be on the lookout for more bones. Altogether 15 skeletal pieces and the skull were recovered. Given the thick brow ridge and hefty, curved thighbone, the bones were first thought to be those of a cave bear, a common find in the region, so not much care was taken in their collection—only the largest pieces were kept. Nevertheless, Carl Fuhlrott, a local schoolteacher and naturalist, was contacted and invited to visit the quarry. When Fuhlrott saw the bones, he identified them as those not of a bear, but of a human.

Fuhlrott noticed, however, that the bones had some unusual features and passed them on to Professor Hermann Schaafhausen at the University of Bonn for a more expert appraisal. Schaafhausen agreed they were not typical human bones and concluded that they represented a natural form not previously seen, even in the "most barbarous races," and that "these remarkable human remains belonged to a period antecedent to the time of the Celts and Germans." Furthermore, Schaafhausen thought "it was beyond doubt that these human relics were traceable to a period at which the latest animals of the Diluvium still existed," by which he meant that this human coexisted with animals such as the cave bear that had gone extinct at the time of some past catastrophic flood.

Being 1857, just two years before Darwin's *On the Origin of Species* would appear, interpretations of fossil remains inspired many theories, and the "Neanderthal" (*thal*, or *tal* in modern German, meaning "valley" or "dale") bones even more so. A leading German pathologist disagreed with Schaafhausen and declared the bones those of a human deformed by rickets. Another scientist rejected both explanations and concluded that the bent leg bones and damaged elbow were just the decades-old remains of a Cossack cavalryman who had been injured in a battle with Napoleon's forces and who crawled into the cave and died.

Darwin's close friend and ally Thomas Huxley took great interest in the "Neanderthal man," and he took some considerable pleasure in poking holes in the skeptics' scenarios. Huxley wondered: Why would a wounded soldier climb 60 feet up a cliff, then remove his clothes and battle gear? And how would he then bury himself under 2 feet of clay, *after he died*? No, Huxley concluded, the Neanderthal man was something different. William King, an Irish geologist, concluded that the Neanderthal man was related to but distinct from modern humans, a separate species, *Homo neanderthalensis*.

Subsequent discoveries in Belgium and France proved beyond any doubt that the skeleton from Feldhofer Cave was no deformed human or Cossack soldier, but that of a distinct human form that was widely distributed across Europe, including as far south as Gibraltar, Spain, and Italy, as far west as England, and as far east as today's Iraq, Iran, and Uzbekistan. Their distinct features—a prominent brow ridge, large nasal cavity, and large, heavy bodies—first inspired popular impressions of Neanderthals as brutes. Such inferior, ape-like, cave-dwelling morons would therefore lie well "beneath" *Homo sapiens*' own scale of self-importance (Figure 12.2).

Over the past 150 years, the cartoonish, science fiction portrayals of Neanderthals have given way to more objective study of Neanderthal history. We have many more, and more complete, specimens of Neanderthals than of any other species beside ourselves, as well as considerable cultural material from a wide range of sites.

Neanderthals occupy a key chapter in the story of modern human origins. It is clear from the fossil record that we *Homo sapiens* and Neanderthals co-existed on the planet for a considerable length of time and cohabited Europe for perhaps 10,000 years before Neanderthals disappeared about 28,000 years ago.

Some key details of the rest of the story have not been clear. The central mystery (from our point of view, of course—it would be a different tale if they were still alive and a Neanderthal were writing this story) is, how are Neanderthals related to *us*?

The questions of what transpired between Nenaderthals and *Homo sapiens* read more like the plot of the TV crime drama *CSI* (*Crime Scene Investigation*) than of highbrow science: What happened that led to the disappearance of Neanderthals, and what role did *Homo sapiens* play? Did we murder and exterminate them? Or, did *Homo sapiens* and Neanderthals meet somewhere for a little romance? How could we solve such mysteries with Neanderthals extinct?

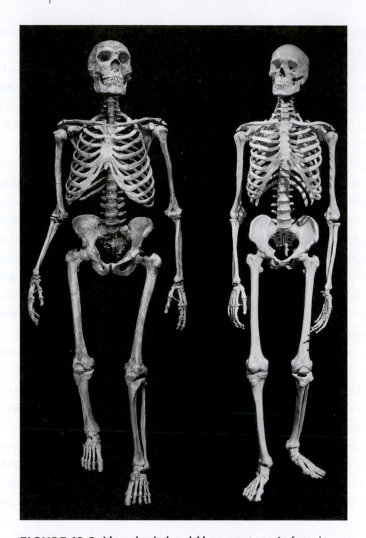

FIGURE 12.2 Neanderthal and *Homo sapiens* At first glance, there is an overall similarity between the two species. But closer inspection reveals that Neanderthals had skulls, chests, fingers, and long bones that differed in shape and thickness from *Homo sapiens*. Denis Finnin © AMNH.

SECRETS OF THE DEAD

Pääbo desperately wanted Neanderthal bone. Being in Germany, he was determined, though, to get not just any Neanderthal's bone, but bone from *the* Neanderthal. Numero uno—the type specimen number 1—that had been found by those quarrymen in 1856. And he didn't want to just see the bone;

he wanted a piece cut out so he could grind it up and extract DNA. That's right, he wanted to pulverize a slice of a national treasure.

He had no idea whether the curators would tell him to get lost. While he thought of how to approach them, he caught a huge break. The curator in Bonn in charge of the specimen *phoned him*. Years earlier, the curator had received requests for Neanderthal samples and had called Pääbo to ask him what he thought the chances of success might be. Pääbo was honest and said perhaps 5%. The curator thought that was too low and denied those requests. Now that methods had improved, the curator was offering bone to Pääbo.

There was discussion about which bone to use. Experience had shown that more compact bones such as arms or legs had more DNA and less contamination than thinner rib bones. Every precaution was taken to further minimize contamination. Protective clothing had to be worn, instruments were treated with hydrochloric acid and rinsed with sterile water, and all of the DNA work was carried out in a lab specially designed for handling archeological specimens and where several measures were used to avoid contamination.

The sample, a half-inch-long, 3.5-gram chunk of the upper right arm bone, was cut out with a sterile drill saw, placed into a sterile tube, and turned over to Pääbo (Figure 12.3).

The bone could reveal whether Neanderthals contributed to Europeans or to any other *Homo sapiens*—if Pääbo's expert team could extract and sequence 42,000-year-old DNA. Bona fide sequences had not been obtained from anything that old except for some mammoths that had been well preserved in permafrost. Because the Neanderthal bones were handled without protection by their discoverers, by museum curators, and by who knows how many more people, contamination of ancient samples with modern human DNA was a terrible worry. Pääbo could not afford to be fooled. The stakes were very high. This was the most ambitious and critical test for ancient DNA. If they screwed up, there might not be a second chance—the fossils were too precious to keep slicing them to pieces.

Pääbo gave the delicate task to Matthias Krings, a graduate student in his lab at the University of Munich. Krings worked in a small room that was kept meticulously clean and irradiated every night to destroy any trace of DNA that might have been left on surfaces. Wearing a protective gown, sterile gloves, a face shield, and a hairnet, he ground the chunk of bone into a powder and extracted the tiny amount of DNA present with a series of solvents. Then, he used the PCR technique to try to amplify short segments of mitochondrial DNA. Finally, the sequences of bases in the amplified DNA were determined.

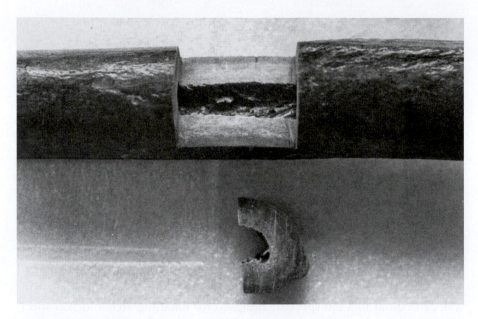

FIGURE 12.3 Getting DNA from the Neanderthal A slice was taken from the upper arm bone of the original Neanderthal type specimen from Feldhofer Cave and processed for DNA extraction and sequencing. Reprinted from Krings et al. (1997). © 1997 with permission of Elsevier.

Pääbo was already in bed and had dozed off when the phone rang.

"It's not human," was all that Krings said.

"I'm coming," Pääbo replied, and he dressed quickly and dashed back to the lab.

The curator from Bonn and Krings were in the lab together, scrutinizing the sequence when Pääbo arrived. The readout was just 61 nucleotides long, but it was obvious that the Neanderthal sequence was very different from those of modern humans.

They cracked open a bottle of champagne as the full impact of their result hit them.

They had done it—they had obtained the first DNA sequence from an extinct human.

The questions and possibilities kept Pääbo awake the whole night. But before he got too carried away, he and they had to be sure. Pääbo wanted the whole procedure repeated, which Krings did. He also wanted to see just how much sequence could be gathered by amplifying adjoining segments of the

Neanderthal DNA. Eventually, the sequence grew to 379 nucleotides. While modern humans vary from one another at an average of 7 nucleotides over this sequence, the Neanderthal sequence differed from that of a modern human at 28 positions.

Still, Pääbo wanted to be as sure as he could be, so before publishing, he wanted to see if another lab could repeat their result. He sent a sample of bone to Mark Stoneking, a former labmate then at the Department of Anthropology at Penn State. Stoneking obtained the same sequences as Krings.

A commentary accompanying the publication of the discovery in 1997 hailed the work as "a landmark discovery . . . arguably the greatest achievement so far in the field of ancient DNA research." For paleoanthropology, it marked another turning point in terms of both practice and theory. Pääbo's team showed that unique genetic information could be extracted from human fossils and that scientists had a whole new tool beyond comparative anatomy and radiometric dating for deciphering human origins.

The big question on everyone's minds, however, was not technical but biological: Did Neanderthals contribute to the *Homo sapiens* gene pool?

Scrutiny of the Neanderthal sequence revealed no evidence that Neanderthals had contributed mtDNA to modern humans. However, that finding did not rule out any interbreeding of Neanderthals and *Homo sapiens*, or any contribution of Neanderthals to the modern human gene pool. The reasons why not come from a bit more consideration of what mtDNA can tell us. Mitochondrial DNA is maternally inherited, so it can reveal possible breeding only between Neanderthal females and modern human males. Furthermore, a given mtDNA lineage can go extinct without a whole species going extinct. It is possible that one Neanderthal had no modern human descendants, but other Neanderthals might have.

Moreover, the inheritance of mtDNA is just one part of the picture. The genes that govern most of our body processes and our anatomy reside in nuclear DNA. Because this DNA exists in only two copies per cell, as opposed to the 100 to 10,000 copies of mtDNA per cell, and because ancient DNA is degraded into very short pieces, the hope for obtaining much, if any, nuclear DNA from Neanderthals seemed very dim at best. In fact, in 1997, when Pääbo's first successes with mtDNA were hailed, the view at the time was that "it is not within our power . . . to retrieve the great majority of lost genetic information."

We are powerless no more.

A SURPRISING RELATIVE
IN OUR FAMILY TREE

Driven by the demand in medical genetics for ever faster and less expensive methods of sequencing human patients' DNA, dramatic advances occurred in DNA sequencing technology in the years after Pääbo first glimpsed a snippet of a Neanderthal sequence. By 2006, Pääbo set his sights on the sequencing not just of some Neanderthal nuclear DNA, but of the Neanderthal *genome*—the entire 3 billion DNA base pairs of an extinct human.

The technical obstacles were formidable. Degradation and chemical decomposition of the DNA leave only short segments of about 50 letters of DNA text. Many tens of millions of these short pieces would need to be sequenced and then assembled in their correct order to reconstruct the genome—a very challenging and complicated task for the best programmers and most powerful computers. And there was the ever present specter of contamination by modern human DNA that could swamp out any signal and confound interpretation.

To decipher the Neanderthal genome and what it could tell us about human history, Pääbo recruited several key scientists with different and complementary expertise. David Reich, a human population geneticist from Harvard Medical School; Nick Patterson, a mathematician and cryptologist from the Broad Institute; Monty Slatkin, a theoretical geneticist from the University of California, Berkeley; Ed Green from his own group at the Max Planck Institute in Leipzig, Germany; and several others formed the Neanderthal Genome Analysis Consortium.

The data dribbled in relatively slowly at first. When the consortium started in 2007, they had about 1.9 million nucleotides of total sequence from three Neanderthal specimens. That nuclear DNA told a story consistent with the previous mtDNA analysis in that no evidence of Neanderthal contribution to the modern human gene pool was detected. But it was too little to rule out that possibility.

Within less than two years, the team had billions of nucleotides of raw sequence. As part of that sequence was being assembled and being compared with the sequence of modern humans, Pääbo received an email from David Reich: "We now have strong evidence that the Neanderthal genome sequence is more closely related to non-Africans than to Africans."

Pääbo was stunned. He understood the implications immediately. If *Homo sapiens* split off from a distant ancestor shared with Neanderthals and never interbred, as Pääbo and many other biologists thought, then all humans should be equally related (or unrelated) to Neanderthals. If Reich was right, the implication was that there had to have been some genetic mixing between Neanderthals and the ancestors of Europeans.

If that were true, it would be a profound discovery about human ancestry.

But if they made that claim and were wrong somehow, it would be a colossal mistake.

They had to be very, very sure.

To look for signs of interbreeding, the team first examined the sequences of five modern humans from different parts of the world: France, West Africa, South Africa, China, and Papua New Guinea. They cataloged about 200,000 positions where the sequence in one individual differed from the sequence in another. Then, they looked at the Neanderthal sequence at these 200,000 positions. If Neanderthals were closer to any group, they should share more of these variable sites. The results were striking. The Neanderthals were not more closely related to either African, but the Neanderthal matched the non-Africans at about 2% more sites. Further scrutiny revealed long segments of DNA in Europeans that matched the Neanderthal sequence. These two lines of evidence clinched it: modern humans picked up some Neanderthal sequences after they migrated out of Africa but before they expanded across the globe.

Where and when did this mixture take place? In order for Neanderthals to contribute to the entire non-African modern human gene pool, the most likely scenario is that interbreeding occurred early, before humans expanded out of the Middle East and across Asia. The fossil record indicates that modern humans appeared in the Middle East before 100,000 years ago, and Neanderthals arrived about 50,000 years ago. Genetic evidence suggests that mixing occurred 47,000 to 65,000 years ago, consistent with the fossil evidence (Figure 12.4).

But the fossil record also indicates that, as modern humans spread, our ancestors replaced Neanderthals entirely as they went, such that Neanderthals vanished completely by 28,000 years ago.

Well, not completely. Pääbo's team estimated that about 1%–4% of non-Africans' DNA is of Neanderthal origin. As Pääbo himself noted, "The Neanderthals live on in many of us today."

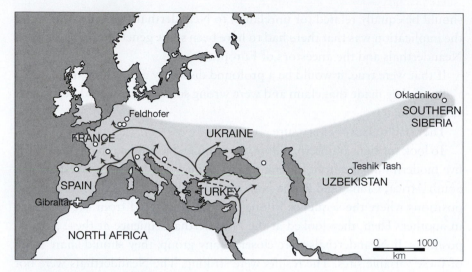

FIGURE 12.4 Neanderthals' Range and the Invasion of Europe by *Homo sapiens* The range of Neanderthals based on fossil and DNA evidence is shaded in gray. Individual sites are shown as open circles and the locations of some key sites are indicated, including the original Feldhofer site and Gibraltar, where the youngest remains have been found. The inferred migration routes of *Homo sapiens* from Africa through the Middle East and into Europe are shown as arrows and dashed arrows.

The figure is based on information in Krause et al. (2007) and Mellars (2006). Map by Leanne Olds.

END-OF-CHAPTER QUESTIONS

1. Why is human contamination a threat to the validity of ancient DNA research on mummies and other human remains? And why is human contamination, even when present, less of a concern for animal tissues?

2. Why is it that mtDNA could show no sign of Neanderthal–*Homo sapiens* interbreeding but nuclear DNA would?

3. Explain the significance of Neanderthals being genetically equally similar to different Africans, but being more similar to Europeans than to Africans.

4. A widely accepted definition of a species is a population of individuals capable of interbreeding. Now that we have evidence of *Homo sapiens* and *Homo neanderthalensis* interbreeding, should we change the definition of our species? Name some other examples of species interbreeding.

Part IV

ECOLOGY

IN 1957, BIOLOGIST AND leading conservationist Julian Huxley said that humans "had been appointed managing director of the biggest business of all, the business of evolution . . . determining the future direction of evolution on this earth." To manage that business, however, requires understanding how nature works, and the factors that influence the diversity and stability of ecosystems. The four stories in this section are about pioneers who discovered important forces that shape ecosystems across the planet. Their discoveries now guide many conservation efforts.

It was a pivotal voyage by another young British biologist, Charles Elton—to a different set of faraway islands than those visited by Darwin and Wallace—that inspired some of the founding ideas of modern ecology (Chapter 13). The testing of ecological ideas in nature was rare, however, until a few naturalists devised experiments that uncovered the surprising roles that individual species can play in shaping communities (Chapter 14), and the crucial relationship between habitat area and the number of species that can be supported (Chapter 15). One of the largest ecological experiments ever undertaken, however, is not intentional. The rising levels of carbon dioxide in the atmosphere discovered by geochemist Charles Keeling (Chapter 16) are the result of the burning of massive amounts of fossil fuels that were buried over eons.

13

YOU ARE WHO YOU EAT

The large fish eat the small fish; the small fish eat the water insects; the water insects eat plants and mud.
—CHARLES ELTON, *ANIMAL ECOLOGY* (1927)

THE *TERNINGEN* PITCHED AND rolled on the heaving, icy Barents Sea, and 21-year-old Oxford University zoology student Charles Elton heaved right along with it. The two-masted schooner had left Tromsø, Norway, two days earlier under the June midnight sun, bound for Bear Island, a desolate, rocky island well above the Arctic Circle, whose most prominent feature was aptly named Mount Misery.

Elton was part of a small advance party of the first so-called Oxford University Expedition to Spitsbergen, a team of 20 students and faculty from various disciplines—ornithology, botany, geology, and zoology—who during the summer of 1921 aimed to carry out an extensive geographic and biological survey of the largest island in the Arctic archipelago northwest of mainland Norway. It was a bold and ambitious undertaking to venture far into storm-tossed and ice-strewn waters, to go ashore and traverse largely uninhabited and partly snow- and ice-covered islands, while risking the whims of some of the most intense weather on the planet—not to mention the fact that *none* of the team had any prior experience in the Arctic. But it was just the sort of adventure Elton and his Oxford colleagues sought (Figure 13.1).

From an early age, Elton had been captivated by wildlife. He spent count-less days walking through the English countryside, watching birds, catching

FIGURE 13.1 Oxford University Spitsbergen Expedition Team (1921) Charles Elton is fifth from the right in the turtleneck sweater; Vincent Summerhayes is to his immediate left. Photo from an account by C. S. Elton written in 1978–1983.

Photo courtesy of Norsk Polarinstitutts Library, Tromsø, Norway.

insects, collecting pond animals, and examining flowers. He began a diary of all of his wanderings and observations at age 13. Elton had been invited on the expedition by his tutor, the eminent zoologist and writer Julian Huxley (grandson of close Darwin ally Thomas Huxley), who was impressed with his student's passion for and command of natural history. Elton had relatively little time to prepare, as his slot on the expedition was confirmed just a month before sailing. His father provided some money, his brother lent him his army boots and other gear, and his mother helped him put together clothes for the two-month voyage. Elton was, by his own account, "very inexperienced, very raw indeed," and had only previously done some rough camping. Huxley encouraged Elton and reassured his parents, "Please assume that there is no danger in the expedition further than that involved, say, in elementary Swiss climbing, certainly much less than difficult mountaineering and indeed negligible."

By just the third day of the voyage, that promise proved empty. The journey to Bear Island was almost 300 miles, and a severe test for Elton, who had never even left England before. The ship was a converted sealing boat whose sleeping quarters previously served to hold blubber, the smell of which could not be removed. That combined with the rough seas spelled utter misery.

Forty-six-year-old medical officer and accomplished mountaineer Tom Longstaff was the most seasoned voyager aboard. Seeing Elton suffering, he said, "I must do something for that poor boy," and proceeded to medicate him with large doses of brandy. With nothing in his stomach, Elton was so drunk when the landing party of seven finally rowed ashore at Walrus Bay on the island's southeast coast, he arrived sitting on top of a load of baggage in the small boat singing at the top of his lungs.

But like Darwin almost a century earlier, Elton's seasickness and other discomforts proved no obstacle to discovery. Charles Elton is nowhere near as famous as Darwin, but this young English naturalist's journey to remote islands also led to encounters with an odd assortment of creatures, presented him with a great mystery, led to an epiphany (or several), and prompted him to write a book that was the foundation of a new field of science. Elton would be known as the founder of modern ecology.

BEHIND THE CURTAIN

Once sober, Elton set up camp with his companions inside the ruins of an old whaling station near the shore, where the ground outside was littered with whale bones, walrus skeletons, a polar bear's skull, and a couple of arctic fox skulls. The plan was for the party to spend a week exploring the southern portion of the 12-mile-long island before rejoining the full team, and sailing on to Spitsbergen, about 120 miles to the north. The four ornithologists in the group were to focus on the birdlife while Elton and botanist Vincent Summerhayes would survey all of the plants and other animals.

After a good night's rest, Elton and Summerhayes went out to explore and collect. Elton dreamed of one day knowing what animals "are doing behind the curtain of cover." Now, he would have a chance to peer into an alien world that few had ever visited, let alone studied.

Situated where the cold polar current meets the Gulf Stream from the west, the island was constantly shrouded in fog, when it was not pounded by sleet or blizzards. Largely flat and dotted with dozens of lakes throughout its interior, and marked by a barren almost moon-like landscape in the north, the island was not without its wonders. On the southern end stood several very tall cliffs overlooking the sea, populated by hundreds of thousands of seabirds. Among the most numerous were black and white common and Brünnich's guillemots, black-legged kittiwakes, and gray fulmar petrels; Elton also saw little auks, Norwegian puffins, and glaucous gulls.

Elton and Summerhayes started their surveys near the camp, then eventually fanned out to various lakes and other features. Elton quickly discovered that he had an excellent, determined partner in the small, slightly built botanist. Just three years older, Summerhayes had seen more of the world, including fighting in the Battle of the Somme (1916). Summerhayes was as keen to identify all of the arctic plants as Elton was to discover the animals, and in the course of their wanderings taught Elton a great deal of botany.

Despite the weather and the shortness of time, the two men were able to survey all of the island's various habitats for plant and animal life. The task was made easier by the relative sparsity of species. For instance, throughout the island, there were no butterflies, moths, beetles, ants, bees, or wasps whatsoever; the insect life consisted mostly of flies and primitive types known as springtails.

Elton had brought various means for collecting and preserving such treasures. He caught flying insects in a butterfly net, and shook other bugs out of plants and leaves onto a white sheet, or uncovered them by flipping over stones. After the insects were killed with cyanide, Elton put them inside tissue paper with labels and packed them into cigar boxes. For aquatic creatures Elton used a series of nets with different mesh sizes. He preserved some aquatic animals inside glass tubes containing alcohol or formalin, which he then corked and sealed with wax.

Elton paid particular notice to what enabled each species to survive in the generally barren and harsh conditions. For example, it was obvious that the seabirds lived off the sea, and that the vast quantities of manure they dropped from the cliffs fertilized the lush vegetation growing below. Other birds Elton found inland such as the snow bunting dined on sawflies, while the arctic skua lived off other birds, stealing their food and eggs.

As the days wore on, and food supplies began to dwindle, the skua weren't the only creatures living off the birds. The ornithologists collected many eggs and birds. The egg shells were preserved by blowing out their contents, which were then made into omelettes. And after skinning, the bird carcasses were put in cooking pots. Elton and the team were surprised at how many species were edible, and he jovially wrote home about having guillemot eggs for breakfast and "stew of fulmar, eider duck, long-tailed duck, purple sandpiper, snow bunting, vegetables & rice!"

Despite storms, running low on bread, being out of margarine for cooking, and having nearly worn through his army boots in just a week, Elton declared life on Bear Island "is rather fun."

FIGURE 13.2 *Terningen* in Ice Fjord of Klaas-Billen Bay, Spitsbergen   From Gordon (1922).

The entire expedition team then headed farther north toward Spitsbergen—and straight into another gale that pushed "growlers," chunks of dangerous, greenish sea ice, into their path. About halfway up the 280-mile-long island, the ship turned toward the azure waters of Ice Fjord, and the clouds parted. The team was treated to a stunning panorama of jagged, snow-capped mountain peaks and shimmering glaciers that crept down pure white valleys to the sea (Figure 13.2).

A whale's spout and then a pod of porpoises broke the glassy surface, while strings of petrels and auks darted just above the water.

More than 200 miles wide, Spitsbergen presented a vast opportunity, and challenge, for exploration. Longstaff helped Elton establish a campsite on the scrabble near a large, several-mile-long lagoon that was half covered with ice, on which many seals lounged. Before returning to the ship, Longstaff cautioned Elton not to walk across the lagoon ice. Elton had learned that Longstaff only gave advice once, and if one forgot it, well, that was one's own mistake.

After nine fruitful days on the island collecting and observing the animals, Elton made that mistake. He was out walking with a geologist on the ice attached to the shoreline of the lagoon. He stepped on a weak spot created by runoff from the shore and broke through into the frigid water. He was saved from going under only by his rucksack, which held him up a bit, and was able to climb carefully back up on the ice. However, freezing cold, momentarily stunned and wearing sun goggles, Elton did not realize that he had been turned around, and almost stepped right back into the hole. Only a shout from the geologist stopped him from another misstep that would have likely drowned him.

Despite being farther north, and more snow and ice covered than Bear Island, summer temperatures were much warmer (some days over 50 °F) and more comfortable on Spitsbergen, at least when it was not snowing. Elton took advantage of the conditions to conduct some actual experiments on the adaptations of arctic animals. To learn more about how they survived in the climate and which habitats they could live in, he tested the tolerance of crustaceans and their eggs to freezing and thawing, and to different concentrations of seawater.

After more than two months of collecting and living on the tundra, the now-veteran Arctic explorer assembled his 33 boxes and bundles of gear and specimens, loaded them all onto the ship himself, and sailed home.

UP AND DOWN AN ARCTIC FOOD CHAIN

Once safely back at Oxford, Elton and Summerhayes assembled the data they had gathered from their surveys. Many naturalists, especially those with a collectors' bent, might have been disappointed at the meager pickings on the Arctic islands. But Elton realized that the relative scarcity of species presented a rare opportunity to describe all of the transactions and relationships among an entire community of organisms—to peel back the curtain on animal life.

While previous naturalists had seen a community as one entity, or as a collection of species, Elton took a novel, functional approach. It was obvious to Elton that in the economy of Arctic islands, the precious commodity was food. So he traced where every creature's food came from.

Food was extremely scarce on land, but it was plentiful in the sea, so he started there. He knew that marine animals (plankton, fish) were eaten by the seabirds and by the seals. The seabirds were eaten in turn by the arctic fox (as well as by the skua and glaucous gull), while the seals were eaten by polar bears. These relationships formed what he dubbed "food chains."

But the connections among the inhabitants of the tundra extended far beyond a few animals. The droppings from the seabirds contained nitrogen, which was used by bacteria, which nourished plants, which produced food for insects, both of which were consumed by land birds (ptarmigan, sandpipers), which in turn were food for the arctic fox. In this manner, the food chains in a community were connected into larger networks that Elton dubbed "food-cycles," later

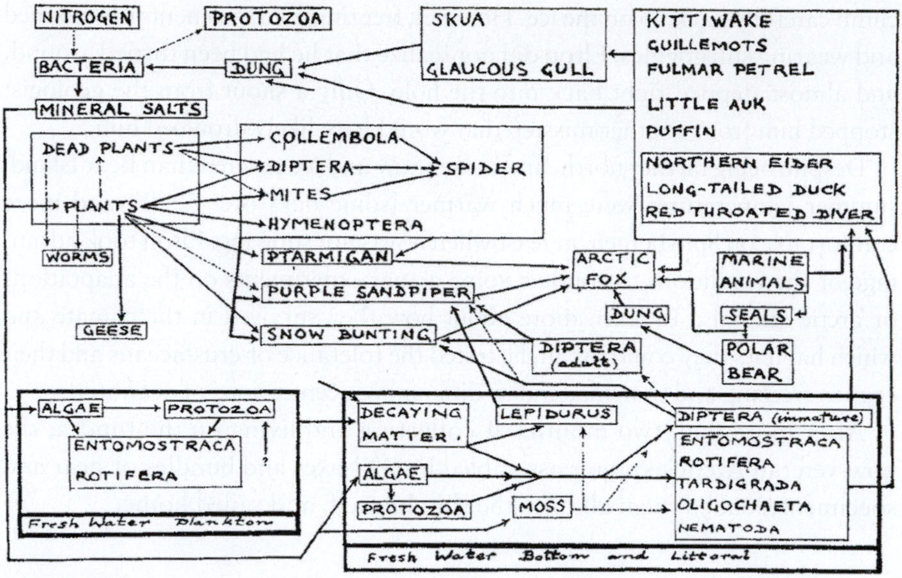

FIGURE 13.3 The Food Chain on Bear Island The first food web drawn by Charles Elton. To follow the food chain on land, start with nitrogen and bacteria at the upper left of the diagram and trace links all the way to the arctic fox. Summerhayes and Elton (1923).

called food "webs." Elton drew a schematic of these chains and webs, the first of their kind, in a paper published with Summerhayes in 1923 (Figure 13.3).

FOLLOWING LEMMINGS

Elton himself moved quickly up the academic food chain, as well as the ranks of the Oxford University expeditions. After receiving his undergraduate degree in 1922, he was appointed to a departmental lectureship in 1923, and named chief scientist for a new expedition to Spitsbergen's little-explored sister island North East Land (Nordaustlandet).

Despite its proximity to Spitsbergen, the island proved much less accessible. The ship was thwarted by thick belts of ice that surrounded the island, and broke its propeller while trying to force its way through a strait. Partially crippled, the ship was pushed about by the ice floes, and the plans to explore the island had to be abandoned.

But for Elton, the journey was not a bust. On the way back to England, the ship stopped as usual at Tromsø. Elton went browsing in a bookstore and

Norwegian Lemmings

FIGURE 13.4 **Lemmings** Lydekker (1904).

happened across a large tome on Norwegian mammals entitled *Norges Pattedyr* by a Robert Collett. Although Elton did not read Norwegian, he was intrigued enough to plunk down one of the remaining three British pounds he had left in his pocket for the journey home for a book that Elton later declared "changed my whole life."

Back in Oxford, Elton obtained a Norwegian dictionary and laboriously worked out a word-by-word translation of one part of the book. There were 50 or so pages on lemmings that, although he had never seen one of the small guinea pig–like animals, enthralled Elton (Figure 13.4). He was captivated by Collett's descriptions of "lemming-years" when the rodents swarmed out of Scandinavian mountainsides and tundra in the autumn in such massive numbers that citizens had taken note of the extraordinary phenomenon for centuries.

Elton drew up charts of the reported events, and found that they occurred with a fairly regular periodicity of three to four years. He made maps and noticed that migrations involving different lemming species in different parts of Scandinavia seemed to occur mostly in the same years. He stared for hours at the maps spread around the floor of his cubicle in an old building at Oxford, thinking there must be something important he was missing. Then, sitting on a seat in the lavatory, it came to him "in a flash." Like Archimedes in his bathtub, Elton on his toilet got the idea that the lemmings were the "overflowing" of a periodically increasing population. At the time, zoologists had assumed that animal numbers largely remained steady. But Elton now realized that they could fluctuate dramatically.

Digging further, Elton wondered how general the phenomenon was. Even though there were no direct reports of lemming migrations in Canada, Elton was able to infer those events by thinking about food chains. He had previously read a book by a Canadian naturalist that had described fluctuations in the populations of other mammals, including the arctic fox. Knowing that the foxes eat Canadian lemmings, Elton located a table of the number of fox skins taken by the fur-trading Hudson Bay Company and, sure enough, found a good correlation between a spike in the number of fox skins and the timing of lemming-years in Norway.

Elton's food chain concept expanded as he realized that lemmings were also preyed on by birds. He also noted that short-eared owls gathered in large numbers in lemming-years in southern Norway, as did peregrine falcons that preyed on the owls.

Lemmings were not the only prey whose numbers fluctuated wildly. Elton learned that the populations of Canadian rabbit also oscillated, spiking enormously about every 10 years before plummeting. The rabbits were prey of the Canadian lynx, and the numbers of lynx (indicated by fur trappers' records) also showed a striking 10-year cycle that correlated with the rabbit cycle.

Elton thought that these correlations offered very valuable glimpses into the workings of animal communities. They demonstrated the remarkable power of populations to increase dramatically. And they showed how the numbers of one species can influence the numbers of other species through food chains.

Elton documented the findings in "Periodic Fluctuations in Numbers of Animals," a 45-page paper he crafted in 1924. He did not know it at the time, but he was laying a cornerstone of the new science of ecology. The young naturalist would next lay the foundation.

LIFE IS ALL ABOUT FOOD

In 1926, Elton's former tutor Julian Huxley was putting together a series of short books on biology. He was keen to have each one written by leading thinkers that would focus on new and emerging principles. Although Elton was just 26, Huxley respected his extensive Arctic experience (three expeditions by that time), and admired the original thinking evident in his published work. Huxley proposed, and Elton accepted to write, a short book on animal ecology.

Elton hurled himself into the project. He wrote like a fiend, most every night from 10 p.m. to 1 a.m. in his flat near Oxford's University Museum, and completed the book in just 85 days! Despite the breakneck pace, the resulting *Animal Ecology* was a classic in terms of both style and substance. The 200-page book had an engaging, conversational tone, with many handy analogies. Its chapters were logically organized around key ideas that introduced and set the agenda for major facets of what Elton saw as the new science of ecology.

Elton explained that his book was "chiefly concerned with what may be called the sociology and economics of animals." The analogy to human society and economics was deliberate. "It is clear that animals are organized into a complex society, as complex and as fascinating to study as human society," and "subject to economic laws," Elton wrote. The connotation was that like human society, animal communities were composed of interacting creatures with different places and roles.

"At first sight we might despair of discovering any general principles regarding animal communities," Elton said. "But careful study of simple communities," as he had performed in the Arctic, "shows that there are several principles which enable us to analyse an animal community into its parts, and in the light of which much of the apparent complication disappears."

Those principles emerged from the central importance Elton placed on food and food chains. He saw food as the "currency" of animal economies. Elton wrote, "The primary driving force of all animals is the necessity of finding the right kind of food and enough of it. Food is the burning question in animal society, and the whole structure and activities of the community are dependent on questions of food-supply."

Food chains formed the "economic" connections among the various members of a community. Chains of animals were linked together by food, and all were ultimately dependent on plants. In Elton's scheme, plant-eating herbivores were "the basic class in animal society" and the carnivores that preyed on them were the next class, and the carnivores that preyed on them the next class, and so on, until one reaches an animal "which has no enemies" at the top of the food chain.

Elton further noted that the animals lower down on a food chain were often abundant, while those at the top were relatively few in number. There was generally a progressive decrease in numbers between the two extremes. Elton called this pattern the "pyramid of numbers."

One example he cited was an English oak wood where one finds "vast numbers of small herbivorous insects like aphids, a large number of spiders and carnivorous ground beetles, a fair number of small warblers, and only one or two hawks." Another example he had documented firsthand was in the Arctic where copious numbers of crustaceans are eaten by fish, the fish eaten by seals, and the seals by a smaller number of polar bears. Elton asserted that such pyramids existed in animal communities "all over the world." Understanding food webs and such pyramids continues to occupy ecologists almost a century later.

Elton also had an enduring, although unintended, impact on popular culture—in fostering the myth of suicidal lemmings. According to Elton's reading of Collett's book, a lemming-year occurred "when a whole lot of lemmings apparently went crazy and walked downhill." He wrote in *Animal Ecology*: "The lemmings march chiefly at night, and may traverse more than a hundred miles of country before reaching the sea, into which they plunge unhesitatingly, and continue to swim until they die." This description, however, was based on yarns collected in Collett's book. Elton had never seen a lemming, nor a migration, let alone a mass suicide.

The myth of lemming suicide received a considerable boost from the 1958 Walt Disney film *White Wilderness*, which depicted lemmings leaping to their demise. After the narrator explained, "A kind of compulsion seizes each tiny rodent and, carried along by an unreasoning hysteria," viewers saw lemmings leaping into the water from a high cliff. The scene was faked—the animals were flung off the cliffs by the filmmakers.

The movie won an Academy Award.

END-OF-CHAPTER QUESTIONS

1. Why did Bear Island provide a great opportunity to decipher entire food chains?
2. Why might there exist a "pyramid of numbers" among animals in a food chain? Which animal was at the top of the pyramid in the Arctic?
3. Elton included a series of proverbs in his book that captured simple truths about the organization of nature. "Large fowl cannot eat small grain" was one saying about the sizes of creatures and the food they ate.

    a. Explain this saying, and offer an example of a predator and its prey.

    b. Can you think of any exceptions to this saying?

4. If lynxes are exterminated by trapping and hunting, what would you predict would happen to rabbit (hare) populations in the short term? In the long term?

14

WHY IS THE WORLD GREEN?

*All animals are equal, but some animals
are more equal than others.*

—GEORGE ORWELL, *ANIMAL FARM* (1945)

IN 1958, MAO ZEDONG, the revolutionary leader of China's Communist Party since the founding of the People's Republic of China in 1949, announced a new five-year plan to industrialize the country's predominantly agrarian economy. During its implementation, dubbed the Great Leap Forward, private land ownership was abolished and agricultural production was organized into large communes made up of thousands of households.

One of the persistent challenges the growing nation of 650 million faced was food shortage. With grain making up to 90% of the Chinese diet, much attention was focused on improving grain production and preventing crop losses. When Mao was told that one of the causes of grain losses was the massive numbers of tree sparrows in the country (a relative of the familiar house sparrow), he ordered the birds to be eradicated from the country, one of four targets of the so-called Four Pests Campaign (the others being mosquitoes, rats, and flies).

The entire nation was drafted into the effort, including children as young as five. Nests and eggs were destroyed, and when the sparrows tried to roost at the end of the day, citizens beat gongs and pots and made them fly until they were exhausted. Hundreds of millions of birds were exterminated.

On the other side of the globe, biologist Bob Paine had a more modest ambition—then again, in comparison to Chairman Mao, everyone on earth had more modest ambitions. Paine wanted to remove a species on a much, much smaller scale. His quest began in a University of Michigan classroom.

BOTTOM-UP OR TOP-DOWN?

One spring day, the sort of day when professors don't feel like teaching and students don't want to be inside, Paine, then a graduate student, was sitting in Professor Fred Smith's class on freshwater invertebrates. The classroom looked out onto a courtyard with a tree that was beginning to bud.

Smith looked out the window and said, "Class, I want you to think about this . . . Why is that tree green?"

"Chlorophyll," answered one student. While technically correct—chlorophyll is the pigment that makes the leaves green—Smith was trying to get the students to think beyond the obvious. It was a teaching style that Paine enjoyed.

"Well, what keeps the leaves there?" Smith replied.

It seemed such a simple question. But, as Paine would learn, the answer was not. Smith was thinking not about chlorophyll, but about food chains. As Charles Elton had shown (Chapter 13), the living world was organized into three main levels according to what creatures they ate, or what ate them. At the base were the producers, the plants that used the sun's energy, water, and soil nutrients to make their own food; above them were the herbivores that ate plants; and finally, above them were predators that ate the herbivores (Figure 14.1).

The general thinking among ecologists at the time was that each level limited the next higher level—that is, the amount of plant food available limited the herbivores, and the number of herbivores limited the number of predators. Thus, the world was organized and regulated from the "bottom" of the food chain up.

But Smith and two close colleagues, Nelston Hairston, Sr., and Lawrence Slobodkin, had some doubts. Herbivores had the potential ability to defoliate trees and consume all of the plants. But the land was generally green, meaning that the herbivores did not consume all of the vegetation. Most plant leaves only showed signs of being partially eaten. To Smith, Hairston, and Slobodkin, these facts meant that herbivores were not limited solely by food; instead something else might be limiting herbivores, namely, predators. They would soon

Bottom Up
Top Down

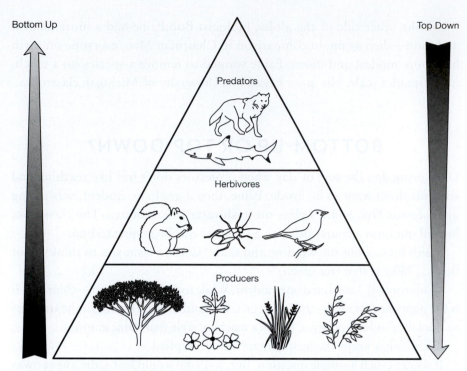

FIGURE 14.1 The Organization of Communities and the Food Chain Three levels of organisms are shown: producers including plants, trees, and algae at the bottom; herbivores that eat producers in the middle; and predators that consume herbivores at the top. The conventional view was that predators were passengers riding atop these communities that were organized from the bottom up. Fred Smith and colleagues challenged that idea and proposed that predators drove the structure of communities from the top down.

Illustration by Leanne Olds.

propose a new idea—that the world was green because herbivores were controlled from the "top" down.

"Why isn't all of its greenery eaten?" Smith continued. "There is a host of insects out there. Maybe something is controlling them?"

Smith's class was the first public airing of what would become known as the green world hypothesis. Flying in the face of the accepted wisdom, the idea drew immediate criticism, the most gentle of which was that it needed testing and evidence.

And Bob Paine would be the first to do so.

THROWING STARFISH

Testing. Paine believed that was what ecology sorely needed. For decades, ecological ideas had been constructed from observation and measurement, but not from experiments like in other branches of biology. After Paine obtained his PhD for his study of small shelly invertebrates called brachiopods, he moved to the Scripps Institute of Oceanography outside San Diego to conduct postdoctoral work. There, he witnessed a debate between two scientists that drew sharp lines between two schools of thought in biology. One scientist argued that description, observation, and measurement were sufficient to understand how nature works. The other, a chemist, said that the way one understands nature is to form a hypothesis, design the equipment to test it, and do experiments. That way, the chemist argued, one knows right away whether one is right or wrong. Paine thought that the latter approach seemed vastly more fun, a great deal quicker, and probably cheaper in the long run.

The structure of the food chain and the bottom-up regulation of higher levels had been taken as givens. There wasn't any interest in testing these ideas, until the green world hypothesis came along and stirred up debate. As that controversy continued unresolved, Paine thought about how he might be able to test it. He needed a system where he could remove the predators and see what happened.

He moved to Seattle in 1962 to become an assistant professor of zoology at the University of Washington, and began to look around the area for a suitable research site. He spent six months searching around Seattle and the university's marine laboratory at Friday Harbor with no luck. His third quarter at the university, he had to teach a class on marine invertebrates, about which he knew almost nothing. He took the class on trips to the muddy bays around Seattle, then to the outer coast of Washington. Paine wound up at Mukkaw Bay, on the tip of the Olympic Peninsula.

There, he later said, he "discovered the Pacific Ocean, and this magnificent array of organisms which lives along its margins." The curved bay faced west into the open ocean, and was dotted with large outcrops. Among the rocks, Paine found a thriving community. The tide pools were full of colorful creatures—green anemones, purple sea urchins, and pink algae, as well as sponges, limpets, and chitons. Along the rock faces, Paine saw that the low tide exposed bands of creatures: small acorn barnacles and large, stalked goose barnacles, beds of black California mussels, and, just below them, some very large, purple and orange starfish called *Pisaster ochraceus*. These bands formed patterns that suggested to Paine that some process was at work that organized

FIGURE 14.2 **Starfish and Mussels on Rockface at Mukkaw Bay** The starfish occur at the lower margin of the mussel bed. Photo courtesy of Robert T. Paine.

the community. He had found his system. "There it was, spread out in front of me, it was nirvana," he later recalled (Figure 14.2).

The next month, June 1963, he made the journey back to Mukkaw from Seattle to start his work. He first needed to understand the predators and prey in the system. There was one small predatory snail, but the main predator in the system was the starfish. To find out what they were eating, Paine would try to sneak up on the animals before they clamped down on the rocks with their arms. Starfish evert their stomachs to engulf and begin digesting their prey. By flipping the starfish over, Paine discovered that they ate a lot of barnacles, but that mussels made up most of their diet.

To find out the starfish's role in the system, he began his experiment. At low tide, he scampered on to a rocky outcrop. He marked out an area 25 feet long by 6 feet tall. Then, with a crowbar in hand, he pried loose all the purple and orange starfish on the slab, grabbed them, and hurled them as far as he could out into the bay.

"It was clearly the dumbest of all experiments, removing starfish to reveal some idea of how their prey would respond, . . . to see how the system worked," Paine later recalled. Dumb or not, twice a month every spring and summer, and once a month in the winter, Paine made the 350-mile round trip journey to Mukkaw to repeat his starfish throwing. On an adjacent plot, he left the starfish alone as a control for his experiment.

By September, just three months after he began removing the starfish, Paine could already see that the community was changing. The acorn barnacles had spread out to occupy 60% to 80% of the available space. But by June of 1964, a year into the experiment, the acorn barnacles were in turn being crowded out by small but rapidly growing goose barnacles and mussels. Moreover, four species of algae had largely disappeared, and the two limpet and two chiton species had abandoned the plot. The removal of the predatory starfish had quickly reduced the diversity of the intertidal community from the original fifteen species to eight.

"I knew that I hit ecological gold," Paine recalled. The results were stunning: one predator controlled the composition of species in a community through its prey—affecting both animals it ate as well as animals and plants that it did not eat.

Paine continued the experiment for the next five years. The mussels advanced down the rock face by an average of almost three feet toward the low tide mark, taking over most of the available space and pushing all other species out. The result revealed that the starfish were necessary for keeping the mussels in check. For the animals and algae of the intertidal zone, the important resource was real estate—space on the rocks. The mussels were very strong competitors for that space, and without the starfish, they took over and forced other species out. The predator stabilized the community by controlling the population of the competitively dominant species.

Paine's results were strong support for the green world hypothesis that predators exerted control from the top down in food chains. But this was just one experiment in one spot on the Pacific Coast. It was crucial to try to replicate the experiment, if possible. Fortunately, Paine soon discovered small, storm-battered, uninhabited Tatoosh Island several miles up the coast from Mukkaw Bay and about half a mile offshore. He found many of the same species clinging to the rocks, including *Pisaster*, and so he began throwing those starfish into the water. Within three months, Tatoosh's mussels started spreading down the starfish-depleted rocks.

Paine had the opportunity to try the same experiment with entirely different species when he took a sabbatical in New Zealand. There, he found a starfish species called *Stichaster australis* that dined on the New Zealand green-lipped mussel, the same species exported to restaurants around the world. Over a period of nine months, Paine removed all of the starfish from one area, and left an adjacent plot alone. He quickly saw clear effects. The area without starfish quickly began to be dominated by mussels, and six of twenty other species initially present vanished in just eight months. Within fifteen months, the majority of space was occupied solely by the mussels.

Paine coined a term to describe the power that such a species had over a community. Drawing on a concept from architecture, he dubbed the predatory starfish "keystone" species, named after the wedge-shaped stone at the apex of a Roman arch. Just like the stone arch that collapses when the keystone is removed, Paine had shown when the keystone species was removed, the ecological community collapsed.

Paine's discovery prompted many questions: Were keystones a general phenomenon, or just a peculiarity of starfish or their habitats? Were any other predators keystone species?

A FORTUNATE ENCOUNTER

As Paine continued his studies a little farther from the shore, another striking pattern drew his attention. In some tide pools, there were a lot of sea urchins and very little kelp (or "seaweed," a form of algae); in other tide pools, there was a lot of kelp and few urchins. Paine suspected that the urchins kept the kelp from growing. So, he and Robert Vadas did another simple experiment: They removed all of the urchins by hand from some pools, or barred them from areas with wire cages. They left nearby pools and areas untouched as controls for their experiment. The removal of the sea urchins had a dramatic effect—several species of algae burst forth in the urchin-free zones. The untouched, control areas with large urchin populations contained very few algae.

The urchins in Paine's pools were eating all of the kelp. So why was nothing controlling the sea urchins? The answer would come from a fortunate encounter far away from his tide pools, on a remote island in Alaska's Aleutian Islands.

In 1971, Paine was asked to visit Amchitka Island. Some students were working on the kelp communities there, and Paine flew out to offer his advice. Jim Estes, a beginning graduate student from the University of Arizona, met with Paine and described his still-developing research plans. Estes was interested in sea otters, but he was not an ecologist. He explained to Paine that he was thinking about studying how the kelp forests supported the thriving sea otter populations.

Paine suggested that Estes might look at the community from a different, top-down perspective. "Have you ever thought about what these animals might be doing to the system?" he asked.

Estes had not thought about the otters in that way. It had not dawned on him that rather than being supported by the kelp forest, the otters might be shaping the system. It was an exciting prospect, so Estes abandoned his plans to study the detailed physiology of the otters—instead he wanted to find out what impact they might have on the kelp forest. Paine did not suggest how Estes might do that, but Estes did see a way.

Estes knew the history of otters in the Pacific, and how the species had nearly been exterminated by the fur trade until it was protected in 1911 by an international treaty. As the animals recovered over the next 60 years, some islands like Amchitka had regained otters, while others had not. So, the "experiment" had been done for him—now what he had to do was to compare islands with otters to islands without the mammals.

With fellow student John Palmisano, Estes traveled to Shemya Island, a 6-square-mile chunk of rock 200 miles to the west without otters. Their first hint that something was very different was when they walked down to the beach and saw huge sea urchin carcasses. But the real shock came when Estes dove under the surface for the first time.

"The most dramatic moment of learning in my life happened in less than a second," he recalled. "And that was sticking my head in the water at Shemya Island. We were in this sea of sea urchins. They were all around us. And there were no kelp anywhere" (Figure 14.3).

The contrast with the kelp forest at Amchitka Island was so dramatic, Estes said, "Any fool would have been able to figure out what was going on."

What Estes and Palmisano figured out was that since otters eat urchins, and urchins eat kelp, when otters were absent, the urchins devoured the kelp. Where otters were present, they controlled the urchin population, and the kelp flourished. The revelation was a powerful demonstration of the green world hypothesis: the predator controlled the herbivore, which in turn allowed the kelp forest to grow.

But the impact of the sea otters reverberated far beyond the growth of the kelp itself. Estes and Palmisano noticed other striking differences between the two communities around each island: colorful rock greenling (a fish), harbor seals, and bald eagles were abundant around Amchitka Island, but not around Shemya Island. The kelp was habitat for many species of fish and all sorts of invertebrates, which were in turn food for seabirds and other animals. An

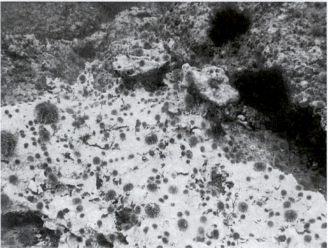

FIGURE 14.3 The Role of Sea Otters in Protecting the Kelp Forest (Top) Where sea otters occur, kelp forests thrive and provide food and habitat for many other creatures. (Bottom) Where sea otters have disappeared, sea urchins devour the kelp, leaving the sea floor barren and depriving other creatures of food and habitat. Photos courtesy of Robert S. Steneck.

entire coastal community depended on the kelp, which depended in turn on the sea otters. The sea otters were clearly another keystone species.

The many indirect effects caused by the presence or absence of the predator revealed previously unsuspected, indeed unimagined, connections among creatures. Who would have thought that otters affected bald eagles? Paine coined a new term to describe the chain of strong indirect effects of keystone species at different levels of the food chain (called trophic levels) in an ecosystem; he called them *trophic cascades*.

A NEW WAY OF SEEING AND MANAGING NATURE

The dramatic and unexpected effects of keystone species and trophic cascades on the Pacific coast raised the possibility that such forces might be operating elsewhere to shape other ecosystems. Over the ensuing decades, as ecologists explored other habitats with new eyes, they have discovered keystone species and trophic cascades across the biosphere in essentially every kind of aquatic and terrestrial habitat—lakes and rivers, coral reefs, marshes, forests, grasslands, deserts, and tundra.

All sorts of predators have been revealed to exert strong top-down effects, from spiders to lions, leopards, and sharks. Once viewed as mere passengers riding atop food chains, many are now known to be drivers of the structure below. But it is also the case that not all predators are keystones, nor are all keystones predators. Herbivores that act as keystones in a variety of ecosystems include bees (through pollination), pika (by sculpting the landscape), and wildebeest (by grazing the savanna).

The revelations, for example, that forests in the Pacific Northwest depend on salmon and bears, or that trees and giraffes in the Serengeti depend on wildebeest, have radically changed our understanding of how nature works. Contrary to earlier notions that every species was important in a system, we now know of many species that have disproportionately large effects on communities. Paine summed up this new view by borrowing a phrase from George Orwell's *Animal Farm*: all animals are equal, but some animals are more equal than others.

The discoveries of keystone species and trophic cascades have also revealed many stupendous mistakes humans have made in trying to manage nature. Mao commanded the extermination of sparrows because he was told that they were eating precious supplies of grain. The next year, the country experienced a massive explosion of locusts that devoured crops. The ensuing food shortage contributed to a famine that claimed many millions of lives. Once it was realized that the sparrows ate not only grain, but insects as well, and thus indirectly protected crops, the campaign against sparrows was halted and redirected against bedbugs.

FIGURE 14.4 Jim Estes and Bob Paine The two pioneer ecologists became lifelong friends. This photo was taken in 2016. Photograph courtesy of Amy Miller (Spine Films).

For many millions of years before humans evolved, large predators roamed the land and seas. But across the globe, we have selectively reduced their numbers to a small fraction of their former strength. We are only now beginning to recognize and reckon with the consequences that reach far beyond the loss of individual species. However, knowledge of keystones and trophic cascades also has the potential to restore degraded ecosystems. For example, the reintroduction of wolves into Yellowstone National Park after a 70-year absence has triggered a trophic cascade that has reduced the overpopulation of elk, and in turn relieved browsing pressures on aspen, cottonwood, and willow, and had far-reaching effects on beavers, fish, and even streams.

Bob Paine did not only identify the first keystone species and coin the concept of trophic cascades (Figure 14.4). In the very last paper he wrote, published on the very day he died at the age of 83, he coined a new term—*hyperkeystone*—to describe the role of humans in affecting every keystone species on earth.

END-OF-CHAPTER QUESTIONS

1. In what way was the green world hypothesis different from the widely accepted view of nature at the time?
2. What was new and original about Bob Paine's approach to studying ecosystems?
3. The island of Rum in Scotland had wolves and trees 500 years ago. It now has neither.
   a. Offer one hypothesis for the current condition based on the concepts of keystone species and trophic cascades.
   b. Suggest one way to test the hypothesis.
4. Draw a trophic cascade that illustrates why Mao's policy of exterminating sparrows backfired.
5. Bob Paine described humans as a hyperkeystone species. In what ways do you think this is true?

15

A RULE OF THUMB
(TO SAVE THE WORLD)

When you have seen one ant, one bird, one tree,
you have not seen them all.

—EDWARD O. WILSON

ONE COLD MARCH DAY in 1954, Harvard professor Philip Darlington called graduate student Ed Wilson into his office.

"How would you like to go to New Guinea?" Darlington asked.

The 24-year-old aspiring naturalist was thrilled. Wilson grew up exploring the swamps, forests, and shores of Alabama, and he long dreamed of seeing the tropics. A journey to New Guinea meant that he could visit and collect on the very islands that Wallace (Chapter 7) had a century earlier! Darlington underscored how little explored the region was, that it was a rare opportunity to be a pioneer.

Wilson knew that Darlington himself was that rare breed of pioneer, and a collector of legendary prowess. Darlington had climbed the entire 6000-meter elevation of the Sierra Nevada de Santa Marta in Colombia, collecting all the way up. He similarly climbed and collected up to the summit of Cuba's and Haiti's highest mountains.

But it was Darlington's own exploits in New Guinea during World War II that revealed his devotion to collecting and science. After the attack on Pearl Harbor (1941), Darlington enlisted in the Army Sanitary Corps Malaria Survey. He served in the military campaigns in New Guinea, the Bismarck Archipelago, the central Philippines, and Luzon, before retiring as a major in 1944.

Before he left New Guinea, Darlington went on collecting trips in several regions of the country, focusing on ground beetles, his primary interest. One day in the jungle, he had crept out on a submerged log floating in a pool to collect water samples when a huge crocodile popped up and swam toward him.

As he tried to move back toward the shore, he slipped off the log into the stagnant water.

The crocodile charged with its mouth open and attacked. Darlington tried to fend him off by grabbing its jaws, but he could not hold on. The crocodile seized Darlington, spun him over and over again in the water, and dragged him to the bottom.

"I was scared and kept thinking: What a hell of a predicament for a naturalist to be in," Darlington later told a reporter. "Those few seconds seemed hours. I kicked but it was like trying to kick in a sea of molasses. My legs seemed heavy as lead and it was hard to force my muscles to respond."

All of a sudden, the croc let go. Despite the muscles and ligaments in his arms being torn, the bones in his right arm being crushed, and both hands being pierced, Darlington swam to the shore, and scrambled up the bank. He headed toward a hospital he knew was nearby—the hike was "the longest I've made," he later said. Once treated, Darlington wrote to his wife and mentioned "an episode with a crocodile" but gave no details.

He needed several months to recover, but his zest for collecting was not diminished. With his right arm useless, Darlington developed a technique for collecting left-handed by using corked vials tied to the end of sticks. He carried the sticks into the jungle, and when he saw his quarry, he jammed the stick into the ground, took out the cork with his left hand, grabbed the specimen, quickly slipped it into the vial, and replaced the cork.

A private and reserved man, Darlington led what his wife described as "an unfragmented life" devoted to science and his family. Wilson and other young naturalists held Darlington in the deepest respect, if not awe. So when Darlington urged Wilson to go to New Guinea while he was "still footloose and fancy-free," it was difficult to resist—even though Wilson was in fact not unattached, but deeply in love and recently engaged to his future wife, Renee. The thought of a long separation at this early stage seemed unbearable, but the couple decided that the expedition was too important to pass up. Wilson would be gone and largely out of contact for 10 months.

STALKING ANTS

Darlington offered Wilson advice on how to make the most of any collecting trip:

> Ed, don't stay on the trails when you collect insects. Most people take it too easy when they go into the field. They follow the trails and work a short distance into the woods. You'll only get some of the species that way. You should walk in a straight line through the forest. Try to go over any barrier you meet. It's hard but that's the best way to collect.

Such guidance might have been wasted on some students, but not Ed Wilson. He took Darlington's words as "those from a master to chosen disciple." Everything about Wilson's upbringing and experience had shown him that the right way was the hard way. There was no privilege in this Harvard student's background. When he was 13 and living in Mobile, Alabama, he took up a newspaper route. That was not unusual for boys of his age, but delivering to 420 customers was. His route was such an enormous job that he had to rise every morning at 3 a.m. just to get home for breakfast by 7:30, and off to school by 8.

Wilson's work ethic was reinforced when he joined the Boy Scouts of America. He thrived on the challenges of learning all about the nature that surrounded Mobile. In just three years, he earned 46 merit badges and rose to the highest rank of Eagle Scout.

Having discovered the pleasure and rewards of hard work, no hardship deterred him from becoming the naturalist he dreamed to be, not even the loss of sight in one eye. The incident happened when he was seven, while spending the summer at a place called Paradise Beach on the Florida panhandle. His parents sent him there to stay with a family friend while they sorted out their failing marriage. Wilson spent every day happily combing the shore and fishing. One day, while fishing for pinfish, a perchlike animal with 10 spines along its dorsal fin, Wilson pulled too hard with a fish on his line and it flew up out of the water and smacked him in the face. One of the spines pierced his right eye. Despite the agony, Wilson kept fishing. The pain subsided over the next days, but the pupil eventually clouded over and he lost vision in the eye.

As a result, Wilson had difficulty spotting animals from far away, such as birds. Fortunately, his remaining eye had unusually acute 20/10 vision so he could see small objects well up close, such as insects. The die was cast: Wilson would become an entomologist. But what kind of entomologist? An article entitled "Stalking Ants: Savage and Civilized" in an old issue of *National Geographic* magazine stoked Wilson's imagination and settled that question. By age 16, Wilson was amassing a collection from his own backyard, making careful descriptions of each species, consulting books and scientific papers, and corresponding with national experts.

Eight years later, ants would be his prime quarry in New Guinea. Wilson would venture alone and without any high technology—all he carried was a hand lens, forceps, specimen vials, notebooks, quinine (for malaria), sulfanilamide (for cuts and infections), and "youth, desire, and unbounded hope."

Wilson's route from Boston took him first to San Francisco, then on to refueling stops at Honolulu and Canton Island before he arrived on Fiji. From there, he went hopping from island to island—New Caledonia, New Hebrides (today Vanuatu), Western Australia, and finally New Guinea (Figure 15.1B). Everywhere he roamed, every species of animal and plant was new to him.

The ants were spectacular—of all sizes, shapes, and colors—and abundant. Wilson was fascinated by red and black forms in one area of New Caledonia that were yellow in another region. He marveled at giant bulldog ants in Australia that were the size of hornets, and easily provoked. In New Guinea, surrounded by the delightful racket of parrots and frogs, he gathered more than 50 species in one spot. Whether steaming in the humid jungle, baking in the dry desert, or hounded by mosquitos, gnats, flies, or leeches, Wilson pushed on. Through dense forest, over steep mountains, into deep valleys, through countless villages, he collected every insect he could find. It was a dream come true, done the hard way.

After the long journey back home—via Sydney, Perth, Sri Lanka, Europe, and New York—it was time to settle back down. The first order of business was marrying Renee, which took place just six weeks after their reunion. The second was to figure out what, if anything, he might have discovered on his Pacific odyssey.

(A)

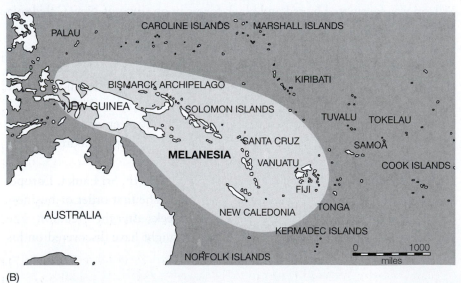

(B)

FIGURE 15.1 Ed Wilson's Melanesian Expedition (A) Ed Wilson crossing a river in New Guinea in 1955. (B) Map of Melanesia. (A) Photo courtesy of E. O. Wilson and Bob Curtis; (B) map by Kate Baldwin.

FINDING PATTERNS

Wilson's main goal had been to collect and classify the ants of Melanesia, at the time a poorly known part of the world. Over the next several years at Harvard, he worked through his collections and described a large number of previously unknown species. The discovery of new creatures was satisfying but Wilson, like Darwin and Wallace before him (Chapter 7), was eager to decipher any patterns that might exist among the distribution of species—patterns that could give him clues to the origins of these island species.

As he collected, Wilson had taken careful notes on where each species lived, their nest sites, their colony sizes, what they ate, and so forth. Wilson recognized that there were in fact two patterns to the distribution of the ants. The first concerned the spread of species across the islands. Wilson figured out that those species that were able to colonize new islands and spread across the chain generally originated in southeastern Asia, and that they favored marginal habitats such as the coast or savanna over the rain forest. When the species made it to another island, they encountered fewer competing species in these marginal areas. Once these immigrants gained a beachhead, some were able to spread inland to occupy and adapt to new kinds of habitat, and eventually form new species.

The second pattern Wilson noticed had to do with the number of species on each island. From his collections and data others had gathered in the Moluccan and Melanesian islands, there appeared to be a correlation between the number of ant species and the overall area of an island. Across the island chains, the number of species increased as island size increased. This relationship was also true within the very large island of New Guinea, where the number of species in a given sector also correlated with area (Figure 15.2A).

Wilson's observations were not new. As early as the 18th century, explorers had noted that larger islands contained more species. The general correlation made common sense—more area usually meant more habitats on an island, and therefore more species occurred that were adapted to those habitats. But when Wilson plotted the logarithm of species numbers against the logarithm of island area, he obtained a straight line (Figure 15.2A). This result suggested a more specific quantitative relationship between species and area.

His mentor Darlington was keenly interested in this relationship, too. While Wilson had been away in Melanesia, Darlington had been quite busy composing a 675-page opus on the geographic distribution of animals. Within an encyclopedic work entitled *Zoogeography*, Darlington noted how area had a profound effect on the number and kinds of animals. He

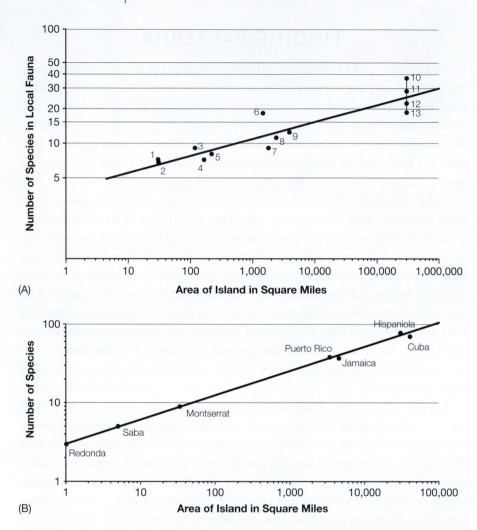

FIGURE 15.2 **The Number of Species Is Related to Area of Habitat** (A) Wilson's logarithmic plot of the increasing number of ant species on 13 islands of increasing area. (B) Plot of Darlington's data on reptile and amphibian species on islands in the West Indies. Both graphs reflect that the number of species approximately doubles with a 10-fold increase in the area of habitat.

(A) From Wilson (1961). Reprinted by permission of the author; (B) from MacArthur and Wilson (1967). Reprinted by permission of the authors.

examined the numbers of reptile and amphibian species on islands of the West Indies including larger islands such as Cuba, Hispaniola, Jamaica, and Puerto Rico, and smaller islands such as Montserrat, Saba, and Redonda (Figure 15.2B). Darlington noted a striking overall relationship: "Division of area by ten divides the amphibian and reptile fauna by two." For example,

Cuba is 10 times larger than Puerto Rico and supports twice as many species. Darlington extended this observation to ground beetles, and suggested that the species–area relationship was a "rule of thumb," meaning not a law such as those of physics, but a correlation that applied often enough to be useful.

That correlation could be stated mathematically as $S = CAz$ where S was the number of species, A was the area of an island, and C and z were constants. For the animal groups Wilson and Darlington had studied, the constant z seemed to fall around 0.3.

FROM A RULE TO A THEORY

For Wilson, his two discoveries of how ant species had spread across the islands, and the relationship between species number and area, spurred yet another idea. Since ant species invaded new islands, and the number of species was constrained by area, then the number of species must be kept in balance—in equilibrium. New species must be replacing existing ones at some rate.

Wilson shared his idea of this dynamic equilibrium with Robert Mac-Arthur, a brilliant young mathematician–naturalist at the University of Pennsylvania. The two men fast became friends and collaborators on island biology. Through many conversations, they sought to come up with more precise models that could account for the patterns of island species.

MacArthur pictured an island filling up with species toward its limit. As the island filled, he realized the rate of immigration of new species would decline because fewer individuals arriving each year would be new forms. At the same time, as the island filled and competition increased, the extinction of existing residents would increase. Thus, each process—immigration and extinction—was a function of the number of species already on the island. MacArthur plotted curves for each process. Where the rates of immigration and extinction balanced one another, the curves intersected and the island was at equilibrium (Figure 15.3).

While encouraged by their mathematical investigations, both ecologists realized that the value of a model or theory was its ability to explain and predict situations other than the ones used to derive it. How could they confirm the rates at which an island fills up when islands across the world are so old that they have reached equilibrium? No one was around to witness their settlement.

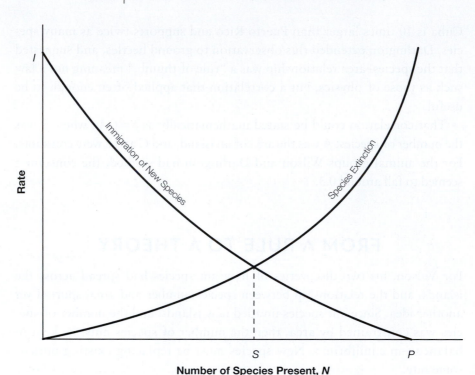

FIGURE 15.3 The Species Equilibrium on Islands The immigration of new species and extinction of existing species are functions of the number of species present on an island. The intersection of those two curves is where the rates balance out and determines the number of species at equilibrium. From Wilson (2010). Reprinted by permission of Princeton University Press.

Wilson realized there was one place where naturalists had witnessed its settlement, or rather, its resettlement: Krakatau.

On August 27, 1883, the small volcanic island between Sumatra and Java was nearly obliterated in one cataclysmic eruption that produced the loudest noise in modern history—heard over 3000 miles away in central Australia. The resulting tsunamis killed over 30,000 people and the massive debris blown into the atmosphere lowered global temperatures by an average of 2 °F over the next year. What was left of the island, as well as two adjacent islands, was buried in ash.

Over the following decades, several naturalists recorded the return of plants and animals to the remaining southern third of the island (now called Rakata). Using their equilibrium model and extrapolating from species–area data for birds on other Asian islands, MacArthur and Wilson calculated that the equilibrium number of birds for Rakata should be around thirty species, and take

40 years to establish, with one new immigrant and one extinction per year thereafter. The data collected by naturalists revealed that by 1908, 25 years after the eruption, thirteen species had recolonized the island; by 1921, 38 years after the eruption, this number rose to twenty-seven, and the number stayed the same through 1934, although five old species went extinct and were replaced by five new species.

The observations were remarkably close to their predictions. Almost exactly the same number of bird species repopulated the island as had lived there before the eruption, although they were not all the same species. Moreover, some species arrived and disappeared during resettlement, confirming that the refilling of the island was a dynamic process.

So far, so good, but Krakatau was just one example. Perhaps MacArthur and Wilson just got lucky that resettlement occurred as it did. There was no way to replicate the study—or was there?

MINI-KRAKATAUS

Wilson fretted over the problem. The destruction of an island was a rare event, occurring perhaps once a century at most. And even if another island did explode somewhere in the world, it might take 30–40 years to observe its resettlement. Wilson could not wait so long for another test of the theory.

Then, it hit him. *Miniaturize the system!* One key inference from MacArthur's curves was that smaller islands would reach equilibrium sooner than larger ones. Instead of studying large islands the size of Krakatau, Wilson realized he should study tiny islands that would recolonize more quickly. They would also have smaller numbers of creatures that could be more readily counted.

Wilson started looking for strings of tiny islands along the Atlantic and Gulf Coasts, and bingo, the Florida Keys jumped out. It had thousands of very small islands. In June 1965, Wilson flew to Miami, rented a 14-foot motorboat, and began exploring the chains of mangrove islets.

Pushing through the shallow mud flats, Wilson quickly found himself alone once again in a wilderness. He drifted among the herons and egrets roosting and hunting in the mangroves. As he explored each patch of "land," which was usually nothing more than a tangle of red mangrove, he collected specimens and took detailed notes.

"No one but a naturalist or escaped convict would choose to traverse the gluelike mud flats or climb through the tangled prop roots and trunks of the mangrove trees," he later wrote. Wilson was that happy naturalist. Eating lunch on the boat, Wilson could peer over the side and see sea squirts, anemones, and small barracuda. "Mine was anything but a world-class voyage, but I was as content as Darwin on the voyage of the HMS *Beagle*."

The tiny islets were rich with species of small creatures—ants, spiders, crickets, caterpillars, and other arthropods. With about 20–50 species each, they were perfect laboratories for his scheme. The next question was how to turn them into mini-Krakatuaus? Wilson considered waiting for hurricanes to drown the little islets, but decided that storms were too infrequent and unpredictable to rely on. The better way to eliminate life was for Wilson to do it himself—by fumigating them.

After Wilson's scouting voyage, graduate student Daniel Simberloff joined the project. A gifted mathematician, but with no field experience, Simberloff would tend the work on the islands while Wilson carried out his duties back at Harvard.

Initial attempts at fumigation with a short-lived insecticide called parathion proved ineffective at killing the insects that burrowed into the trees. Wilson and Simberloff asked their Miami-based exterminator about other options. Methyl bromide, a poisonous gas, was often used to fumigate houses, and eliminated termites that had burrowed into the woodwork. Performing the operation on an islet surrounded by water would be tricky, but the resourceful exterminator was up for the challenge.

In October 1966, the extermination crew built a scaffold and enclosed the first islet in a tent, then pumped in the gas (Figure 15.4). The next day, Wilson and Simberloff scoured the islet for any sign of life and found none. Having worked beautifully, the procedure was repeated on six islets; two nearby islands were left untouched as controls.

Once the tenting was removed, the empty islands were available to any colonizing creatures that might happen by. Simberloff stayed in the area for a year and took surveys every 18 days to monitor the islets.

He saw life return quickly. Within weeks moths, bark lice, and ant queens had landed. Spiders also arrived by ballooning from nearby islands attached to silk threads carried by the winds. Within eight months, all of the islands except the one most distant from large, untreated islands had regained similar numbers and kinds of species as before their fumigation, although not exactly the same set of species. In the course of that time, and as predicted by MacArthur and Wilson, many potential colonists were observed to arrive and disappear. The settlement of the empty islands was a dynamic process.

FIGURE 15.4 The Fumigation of a Mangrove Islet (1966) Daniel Simberloff is in the foreground, and one of the experimental islets to be fumigated is in the background.

From Wilson (2010). Courtesy E. O. Wilson and Steve Tendrich.

Simberloff later tested the effect of area on colonization by fumigating a set of larger mangrove islets, allowing them to be recolonized and equilibrate, and then removing sections of several islets with chain saws and handsaws. As predicted, the smaller, reduced islets supported fewer species than their larger parents. The dynamics of species turnover on the shrunken islets demonstrated that species are more likely to go extinct on smaller islands.

FROM THEORY TO THE REAL WORLD

Wilson and MacArthur published *The Theory of Island Biogeography* in 1967. It was a small book full of graphs, equations, mathematical reasoning, and data about island species. But their perspective was not limited to the mathematics of life on remote islands. On the very first page, they wrote:

> Many of the principles graphically displayed in the Galápagos Islands and other remote archipelagoes apply in lesser or greater degree to all natural

habitats. Consider, for example, the insular nature of streams, caves, . . . tide pools, taiga as it breaks up into tundra, and tundra as it breaks up in taiga. The same principles apply, and will apply to an accelerating extent in the future, to formerly continuous habitats now being broken up by the encroachment of civilization.

The generalization from islands to other isolated pieces of nature was grasped immediately by fellow ecologists and conservationists. The relationships between island area and the number of species that could be supported, and effects on extinction rates, had potentially profound implications for the design and effectiveness of parks and reserves. Such places were generally islands of nature in seas of human-sculpted landscapes. The theory of island biogeography inspired a flurry of research and debate about the sizes and shapes of reserves necessary to preserve species, and the impacts of roads and other barriers on the species within them.

In the 50 years following his first major work, Ed Wilson became one of the most honored biologists of his time. He went on to found another field of biology: sociobiology, an outgrowth of his pioneering work on social insects. An exceptionally gifted and prolific writer of more than 25 books, Wilson earned two Pulitzer Prizes for writing about science, as well as the National Medal of Science.

These are exceptional achievements for any scientist (or writer), but Wilson has also used his eloquence and deep knowledge of biodiversity to draw greater attention to the accelerating impact of human activity on nature. As species populations and ranges have dwindled, with many reaching critically endangered status, Wilson's calls have become increasingly urgent and bold.

In his book *Half-Earth*, Wilson proposed that a new goal should be set for conservation: to dedicate half the surface of the earth (land and water) to the safeguarding of nature. He pointed out that all conservation efforts over recent decades, no matter how noble, were piecemeal and had been insufficient to stem the threat of a major extinction of species. Therefore, a bolder approach was needed, one that was based on the species–area relationship he first encountered almost 60 years earlier: "With the relation of sustainable species to the area of the habitat . . . the fraction protected in one-half the global surface is about 85%." Wilson further pointed out that the fraction could be increased by including within the one-half various "hot spots" that included greater biodiversity or endangered species.

With about 15% of the land surface and about 3% of the oceans now protected, there is a long way to go. But Wilson's Half-Earth proposal has inspired many organizations and even governments to aim higher. Hopeful that humanity will change course in time, Wilson still sees great potential: "Earth, by the 22nd century, can be turned, if we so wish, into a permanent paradise for human beings, or at least the strong beginnings of one."

END-OF-CHAPTER QUESTIONS

1. Describe the two patterns Wilson observed regarding the distribution of ants on Melanesian islands.
2. Explain the factors that led Wilson to the idea of (a) studying small islets, (b) fumigating them, and (c) observing the recolonization of species.
3. According to the theory of island biogeography, why might one large nature preserve be more effective than several smaller preserves (of equal total area) at safeguarding species?
4. How might the construction of roads impact the number of species that persist within a wilderness?
5. What characteristics did Ed Wilson display at an early age that contributed to his success as a naturalist? Which of these traits are shared with Charles Darwin (Chapter 7), Alfred Wallace (Chapter 7), Mary Leakey (Chapter 11), Charles Elton (Chapter 13), or Robert Paine (Chapter 14)?

16

THE FUTURE OF LIFE ON A HOCKEY STICK

*There is a great danger in refusing to believe things
you do not like.*
—WINSTON CHURCHILL

A "PLAN" WOULD BE too strong a word for Charles David Keeling's notion of what to do after getting his PhD in chemistry.

In 1953, just eight years after World War II, the demand for chemists among oil companies and large chemical manufacturers was very high. Like his fellow graduate students, Keeling—known as Dave to his family and friends—had many options and several offers. But unlike his colleagues, Keeling was not motivated by a well-paying industry job as a polymer chemist. He wanted instead to pursue his budding interest in geology, which meant becoming a low-paid postdoctoral fellow.

A passionate outdoorsman who was drawn to mountains and glaciers, Keeling also decided that he did not want to work in the eastern United States. So he took the rather bold approach of writing to every geology department west of the North American continental divide, and offering his services as a chemist. Nearly all declined, except Professor Harrison Brown of the California Institute of Technology, who had recently launched a new department of geochemistry. Brown invited Keeling to Caltech to be his first postdoctoral fellow.

Once he arrived in Pasadena, Keeling needed to find a research project. He was being paid through a grant from the Atomic Energy Commission for researching the extraction of uranium from rock. The project involved a lot of rock crushing, which Keeling did not find appealing; he wanted to apply his chemistry skills to more interesting geological problems.

One day, he heard Brown speculate about one such problem. Rivers and groundwater often pass through or over limestone, which is made mostly of calcium carbonate. Some carbonate dissolves in the water and affects its qualities for household or agricultural uses. The problem that interested Brown was what determined the amount of carbonate in water. His hypothesis was that the carbonate in water should be in equilibrium with the limestone and atmospheric carbon dioxide.

Intrigued by the practical nature of the problem, and excited by the realization that working on it would require him to go outdoors into the field, Keeling asked for the chance to test the idea. Brown agreed. Little did Keeling know that his first foray into geochemistry would shape the next 40 years of his life, and lead to truly earthshaking revelations.

THE FOREST BREATHES

The project seemed straightforward: to measure and compare carbonate in water and carbon dioxide in air. Keeling liked to design and build things. He quickly assembled a system to extract and measure carbonate from water. Through a series of steps, the dissolved carbonate was converted to CO_2 gas. The amount of gas was then measured in a manometer, a device with a confined space above a column of mercury whose height varied according to the amount of gas present. Keeling determined that his modified design was as or more accurate than any previous instrument—within 0.1% accuracy—and he was eager to give it a try in the field.

Before setting out, however, Keeling learned that the amount of carbon dioxide in the atmosphere was not well known. He could not just rely on published values because they varied widely, from 150 to 350 parts per million (ppm). He would need to directly measure CO_2 in the air. There did not appear to be a standard way to do that, so Keeling tackled that problem, too. To collect air samples, he had a dozen 5-liter glass flasks constructed that were each closed off with a stopcock to hold a tight vacuum. He collected samples by pointing the flask into the wind, then opening and closing the stopcock (Figure 16.1). Afterward he extracted the CO_2 from the air sample, and measured it in his manometer. In test samples Keeling took around Pasadena, he found variable amounts of CO_2. It was obvious that emissions from cars,

FIGURE 16.1 Charles Keeling and His Flask

Special Collections and Archives, UC–San Diego; National Geographic Image Collection/Alamy Stock Photo.

industry, and backyard incinerators influenced his measurements. The city was not a good place for studying the atmosphere, so he looked for more pristine settings.

The iconic Big Sur area on the California coast was a good prospect. It was about a day's drive from Pasadena, and the groundwater there contacted limestone rocks. Keeling set up camp in Big Sur State Park for three weeks. He was not sure whether even in the fresh air along the coast the CO_2 levels were constant, so he decided to take air samples every few hours, which meant climbing out of his sleeping bag in the middle of the night. He also took numerous water samples.

Back in the lab at Caltech, Keeling's measurements revealed that Brown's hypothesis was not correct. The river and groundwater carried more carbonate than the atmosphere.

That hypothesis was soon forgotten as Keeling detected potential signs of a much more interesting phenomenon. The air samples taken at night contained markedly more CO_2 than those observed during the day, a difference as large as 70 ppm. One possible reason for the difference emerged when Keeling measured the ratio of carbon isotopes in his CO_2 samples. It was known that plants preferentially utilize carbon-12 over carbon-13 during photosynthesis. The nighttime samples were reduced in carbon-13 relative to carbon-12. Keeling thought the difference in the isotopes might indicate that the rise in CO_2 during the night was due to its release from plants and soil.

Was the forest breathing in CO_2 during the day and releasing it at night?

To find out, Keeling had to check many more locations far from cities, which suited him perfectly. Newly married, he and his wife, Louise, took data collecting trips that became joint adventures to wildernesses all over the west coast of the United States—to Yosemite National Park, redwood forests, the Sierra Nevada range, the Anza-Borrego desert, the high mountains of Arizona, and Olympic National Park in Washington State.

Everywhere the Keelings went, Dave found the diurnal pattern of lower CO_2 levels in the day and higher levels at night. Moreover, whether in a desert or a rain forest, near the sea or on a high mountain, the level of CO_2 in the afternoons was very similar: about 310 ppm. That consistency suggested that while local biology produced a daily cycle of CO_2, the heating and mixing of the air during the daytime produced a near-homogeneous concentration of CO_2 throughout the atmosphere.

In a short time, Keeling had brought precision and clarity to the study of atmospheric CO_2. His wide sampling and consistent methods revealed both daily dynamics and an overall constancy in CO_2 levels that previous scientists had largely missed, and he now understood the reasons for the disparities among previous efforts.

A VAST GEOPHYSICAL EXPERIMENT

It was widely appreciated that even though CO_2 is not as abundant as other chemicals, being just 0.03% of the atmosphere by volume, it is profoundly important to the biology, geology, and climate of the planet. Life depends on CO_2 as its primary source of carbon (via photosynthesis), the molecule plays an important role in the weathering of rocks due to its acidic properties, and it had been known for a century that CO_2 is an important *greenhouse gas* that strongly affects the heat balance in the atmosphere. Many scientists were interested in these roles of CO_2, so Keeling and his work caught their attention.

Harry Wexler, the director of research at the U.S. Weather Bureau, was one who took notice. Wexler was intrigued that the levels of CO_2 in the atmosphere might be far less variable than had been generally thought. He invited Keeling to Washington, D.C., for a chat.

After the long plane trip—Keeling's first ever—he was ushered into the old, crowded quarters of the Weather Bureau. Wexler explained that the following year (1957) was being touted as the International Geophysical Year, and that various countries and U.S. government departments were launching projects to mark the event. The Weather Bureau was planning to measure atmospheric CO_2 at several remote locations, including the top of Mauna Loa volcano on the island of Hawaii. Wexler was impressed that Keeling had successfully measured CO_2 at high altitudes and asked for his input.

Keeling expressed concern that the project would fail if measurements were taken by the same methods that he had shown to be unreliable. He suggested that his flask technique could be used to calibrate a set of automated gas analyzers deployed around the world. Wexler agreed with Keeling's advice and asked him whether he would like to move to Washington to carry out the project.

As Keeling considered Wexler's offer, another scientist heard about his work. Roger Revelle was the director of the Scripps Institution of Oceanography outside San Diego, and was keenly interested in the role of the oceans in exchanging CO_2 with the atmosphere. He had become particularly interested in the role of the oceans in absorbing the additional CO_2 that was being produced by the burning of fossil fuels—the oil, gas, and coal formed over millions of years from buried, decayed organisms.

Revelle was not the first scientist to consider the possible long-term effect of the consumption of fossil fuels and the release of additional CO_2 into the atmosphere. In the late 19th century Swedish chemist Svante Arrhenius recognized that burning fossil fuels added CO_2 to what he called the atmospheric "hothouse," although he did not foresee a noticeable impact for centuries or more. In the 1930s, Guy Callendar, an English coal engineer, calculated that millions of tons of carbon dioxide had been added to the atmosphere and gathered temperature records that indicated the world had warmed by about 1 °F in the previous 50 years.

Revelle and his Scripps colleague Hans Suess thought at first that most or all of the additional CO_2 would be absorbed by the oceans within about a decade of its release. But then they recognized that the chemistry of the ocean was such that much of the absorbed CO_2 would be exchanged back into the atmosphere. In 1957 they wrote:

> Thus human beings are carrying out a large scale geophysical experiment of a kind that could not have happened in the past nor be reproduced in the future. Within a few centuries we are returning to the atmosphere and oceans the

concentrated organic carbon stored in sedimentary rocks over hundreds of millions of years.

They cautioned:

> The increase of atmospheric CO$_2$ from this cause is at present small but may become significant during future decades if industrial fuel consumption continues to rise exponentially.

Revelle invited Keeling to visit Scripps to discuss the possibility of a job for him there. They had lunch together in Hans Suess' backyard on a typical brilliant, sunny day in San Diego, with a gentle breeze wafting in from the sea. The prospect of a dim basement office in Washington was no match for such a setting, so Keeling decided to move to Scripps.

Revelle's idea was to take a snapshot of CO$_2$ levels around the globe to establish a baseline. Then, perhaps 20 years later, someone else would repeat the measurements to see if CO$_2$ levels had changed.

Keeling, however, had a different idea. Experience had taught him that frequent measurement and great accuracy were crucial to sorting out spurious readings, and could also reveal dynamics that single measurements could not. He advocated for continuous monitoring, a technically challenging, logistically complex, and more expensive program.

Fortunately, Wexler at the Weather Bureau was gracious about Keeling's decision to work at Scripps, and still wanted him involved in his agency's monitoring project. So Wexler provided funding for Keeling's salary at Scripps and for major equipment, as well as access to remote weather stations. Revelle was eager to measure CO$_2$ over the oceans and throughout the atmosphere, so a plan was developed to measure CO$_2$ continuously at the weather stations on Mauna Loa in Hawaii and at Little America in the Antarctic, as well as aboard a ship at sea and from a military plane aloft.

THE PLANET BREATHES, WHILE CO$_2$ RISES

Up to this point, all of Keeling's measurements had been obtained using the flasks he designed. They had worked well and were inexpensive, but each individual measurement was time-consuming. Larger-scale studies would require a new automated approach. Keeling worked diligently on calibrating several automated gas analyzers using his flask technique before they were shipped

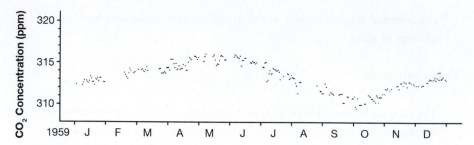

FIGURE 16.2 The Planet Breathes The seasonal variation in atmospheric CO_2 at Mauna Loa. Keeling interpreted the rise in late fall and the decline in the spring as being due to the release and uptake, respectively, of CO_2 by plants in the northern hemisphere. Data are reported as the number of molecules of carbon dioxide divided by the number of molecules of dry air multiplied by 1 million (ppm).

out to be installed. Once installed, however, operating them in remote places where electrical power was unreliable proved challenging. It took almost a year and a half, until the summer of 1958, before both the Mauna Loa and Antarctic machines were up and running.

The principal task was to accurately measure CO_2 levels in different parts of the world on a continuous basis. The goals were to find out, first, whether such monitoring was technically feasible and, second, how values compared in different parts of the world.

Soon after Keeling visited Mauna Loa in November of 1958, however, the instrument detected some unexpected variation. The CO_2 level on the volcano gradually rose month by month. Then, beginning the following May, the level started to decline, before rising again in late October and early November (Figure 16.2). Keeling realized that the maximum CO_2 level on Mauna Loa occurred just before plants in the northern hemisphere put on new leaves. "We were witnessing for the first time nature's withdrawing CO_2 from the air for plant growth during the summer and returning it each succeeding winter," he wrote.

The data from the Antarctic told a different and very surprising story. There was no pronounced seasonal fluctuation as in the northern hemisphere, probably because of a smaller area of growing plants in the temperate and polar regions of the southern hemisphere. There was, however, a small but steady rise in CO_2 levels such that over just two years, CO_2 levels rose by almost 3 ppm or 1% (Figure 16.3). Scrutiny of the Mauna Loa data revealed a similar year over year rise, on top of the seasonal variation.

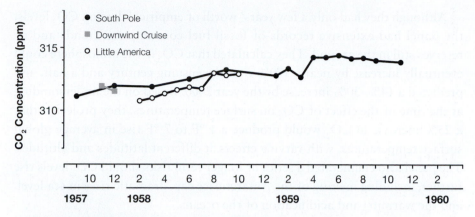

FIGURE 16.3 The First Sign of Rising CO$_2$ The rise in CO$_2$ in the atmosphere was first detected over the Antarctic after just two years of measurements. Data are reported as the number of molecules of carbon dioxide divided by the number of molecules of dry air multiplied by 1 million (ppm).

No one had expected to detect a long-term rise in CO$_2$ levels over the short time span of the initial study. The discovery that CO$_2$ levels might be rising so quickly was astonishing and alarming. The decision was made to continue monitoring at multiple locations. Over the next several years, the upward global trend was confirmed.

IMPLICATIONS AND REACTIONS

Even though the overall change in CO$_2$ observed was relatively small, and the finding published in a technical journal, news of Keeling's discovery reached policy makers—and some paid attention. As early as 1965, in a special message to Congress, President Lyndon Johnson said, "This generation has altered the composition of the atmosphere through radioactive materials and a steady increase in carbon dioxide from the burning of fossil fuels."

Johnson commissioned a blue-ribbon scientific advisory committee to prepare a report that analyzed a variety of threats to the environment. Revelle and Keeling served on a panel that wrote a detailed appendix entitled "Carbon Dioxide from Fossil Fuels—The Invisible Pollutant" in which the connection between the burning of fossil fuels and the rise in atmospheric CO$_2$, and the potential climatic consequences, were explored and explained.

Although they had only a few years' worth of empirical data on CO_2 levels, the panel had extensive records of fossil fuel consumption trends and of reserves still in the ground. They calculated that CO_2 in the atmosphere could eventually increase by nearly 170% over the ensuing century and a half, and predicted a 14%–30% increase by the year 2000. Based on the understanding at the time of the effect of CO_2 on surface temperatures, they projected that a 25% increase in CO_2 would produce a 1 °F to 7 °F rise in average global surface temperatures, with varying effects at different latitudes and altitudes. The panel also enumerated a series of longer-term risks should CO_2 levels rise further, including melting of the Antarctic ice cap, dramatic rises in sea level, and the warming and acidification of the oceans.

Johnson applauded the committee's work and stated, "This report will surely provide the basis for action on many fronts."

POSSIBILITY BECOMES REALITY: TEMPERATURES RISING

Keeling himself avoided politics and activism, and focused on obtaining and interpreting more data. What he and the president's advisory committee had raised was the *possibility* of global warming due to increasing CO_2. Only more monitoring of the global environment could tell what was actually happening and how quickly.

For the next 40 years, Keeling worked tirelessly to ensure that CO_2 monitoring continued. He managed to navigate through numerous changes in personnel and priorities within several government agencies to keep the program running. By 2000, CO_2 levels had risen to 369 ppm, 15% higher than and within the range predicted by Keeling and his fellow panelists in 1965. And the average global surface temperature had risen by a little less than 1 °F, just below the low end of their estimate 35 years earlier. When plotted over the prior 1000 years of earth history, the graph of the recent rapid rise in temperature was likened to the shape of a hockey stick, a long shaft of relatively stable temperatures with a sharp upturned blade at the end, rising further to an unknown future level (Figure 16.4).

That graph and the plot of rising CO_2 levels known as the *Keeling curve* (Figure 16.5) have become iconic symbols of the reality of global warming and climate change. A subsequent director of Scripps stated that Keeling's records of carbon dioxide levels "are the single most important environmental data set taken in the 20th century," adding, "Dave Keeling was living proof that a

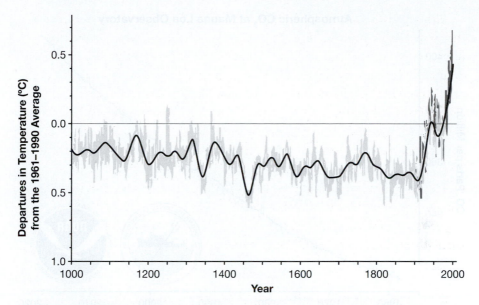

FIGURE 16.4 The Hockey Stick Graph of Earth Temperatures, AD 1000–2000
Reconstruction of northern hemisphere surface temperature variations over the past millennium based on the work by Mann et al. (1999).

Adapted from Figure 1(b) from the Summary for Policymakers, Intergovernmental Panel on Climate Change (2001). Reprinted with permission.

scientist could, by sticking close to his laboratory bench, change the world."
In 2002, President George W. Bush presented Keeling with the National Medal of Science: "For his pioneering and fundamental research on atmospheric and oceanic carbon dioxide, the basis for understanding global carbon cycle and global warming."

After Keeling passed away in 2005, his son Ralph, also a Scripps scientist, took over the CO_2 monitoring program. By the end of 2016, atmospheric CO_2 had climbed to 405 ppm (Figure 16.5). Based on the study of ice cores, this is about 45% higher than preindustrial levels (around 280 ppm), and 100 ppm higher than at any time in the prior 800,000 years of the earth's history. The last time global CO_2 was 400 ppm was 3–5 million years ago. Moreover, the rate of increase has accelerated from about 0.7 ppm/year when Keeling first started his measurements to about 2.1 ppm/year over the past decade. That rate is more than 100 times faster than when the last ice age ended.

The year 2016 was also the warmest year in a 136-year record reaching back to 1880 during which time the average surface temperature has increased by

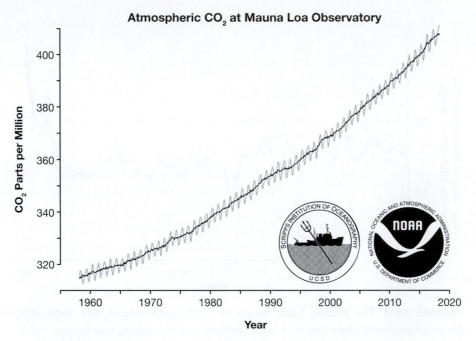

FIGURE 16.5 **The Rising Level of CO$_2$ Measured on Mauna Loa, Hawaii** The carbon dioxide readings on Mauna Loa. The oscillating plot reflects the seasonal variation on the mountain. The smooth curve represents the seasonally corrected data. Data are reported as the number of molecules of carbon dioxide divided by the number of molecules of dry air multiplied by 1 million (ppm).

Adapted from National Oceanic and Atmospheric Administration (2018); NOAA Earth System Research Laboratory, Global Monitoring Division.

about 1.7 °F. Indeed, 16 of the 17 warmest years on record have occurred since 2001.

Many countries have endorsed the goal of limiting global warming to 2 °C, or 3.6 °F, expressing a collective sense that the potential impacts of additional warming are too great to risk. At present, the attainability of that goal is uncertain in light of the continued increase in the consumption of fossil fuels, which is now three times greater than when CO$_2$ monitoring began in 1958. And thus, the future climate on the earth—the blade of the hockey stick—and all of the many ramifications for the planet's inhabitants, human and otherwise, are uncertain.

END-OF-CHAPTER QUESTIONS

1. Why were reports of atmospheric measurements of CO_2 taken before Keeling widely variable and often inaccurate? How did Keeling improve the accuracy of measuring CO_2?
2. What discoveries did continuous monitoring of CO_2 enable, as opposed to taking "snapshots" of CO_2 levels many years apart?
3. What is the significance of the accuracy of Keeling's and Revelle's predictions made in 1965 about the future increases in CO_2 and temperature?
4. On the eve of a war that the British and other people hoped would not break out, Winston Churchill warned, "There is a great danger in refusing to believe things you do not like." Discuss the psychological and societal factors that have contributed to the denial of, and delayed action on, climate change.

Part V

PHYSIOLOGY AND MEDICINE

HUMAN AND ANIMAL BODIES carry out many complex, indeed, amazing tasks. For example, all of the cells and tissues of the body develop from just a single fertilized egg cell. The immune system defends against millions of potential microbial enemies. And our brains generate speech, language, music, art, and ideas. These organs, systems, and processes have posed some of the greatest mysteries in modern biology. The three stories in this section reveal how the mysteries of development (Chapter 17), immunity (Chapter 18), and the brain (Chapter 19) were penetrated by a few seminal experiments. Each breakthrough was recognized by the Nobel Prize in Physiology or Medicine.

17

UNLEASHING POTENTIAL

The mind is not a vessel to be filled,
but a fire to be kindled.
—PLUTARCH

"DRAW AN ORANGE," THE teacher said.

It was part of an intelligence test for school. But being 1941, the second year of Britain's fight for survival in World War II, with food and other necessities strictly rationed, oranges or bananas were luxuries that eight-year-old John Gurdon would not see until well after the war. Nevertheless, the boy started to draw what he thought an orange looked like. Reasoning that the fruit would not just exist in space, he started drawing a stalk by which it would hang from a tree.

The teacher grabbed the piece of paper and tore it up, and later told Gurdon's parents that he was "mentally subnormal" and would need special teaching. His parents instead promptly moved him to another school, where he thrived and spent his spare time collecting moths and butterflies and raising their caterpillars.

A few years later, his parents enrolled Gurdon at Eton, the very prestigious boarding school founded by King Henry VI in 1440, where John continued his insect hobby. His science teacher assessed his talent and prospects in a report to his parents:

It has been a disastrous half. His work has been far from satisfactory. His prepared stuff has been badly learnt, and several of his test pieces have been torn over; one of such pieces of prepared work scored 2 marks out of a possible 50. His other work has been equally bad, and several times he has been in trouble, because he will not listen, but will insist on doing his own work in his own way. I believe he has ideas about becoming a Scientist; on his present showing this is quite ridiculous, if he can't learn simple Biological facts he would have no chance of doing the work of a Specialist, and it would be sheer waste of time, both on his part, and of those who have to teach him.

Gurdon's first science course at Eton would be his last. After this report from his biology master, the 15-year-old was not allowed to take further science courses and spent the next three years studying ancient Greek, Latin, and other subjects thought fit for the bottom rung of students.

The biology master was not among Gurdon's guests when, many years later, King Carl XVI Gustaf of Sweden awarded Gurdon the Nobel Prize.

FINDING A PATH

How did such a seemingly terrible student, with no basic science education to speak of, manage such a feat? Not easily.

Despite his experience at Eton, Gurdon was determined to pursue science. His narrow education in classics, however, made entrance to college difficult.

Fortunately, Gurdon's family had some connections. Indeed, the Gurdon lineage had very deep roots in England. A Gurdon ancestor had come to England from Normandy in 1066 with William the Conqueror. One of his descendants, Bertram de Gurdon, was said to have fired the crossbow that fatally wounded King Richard I (the Lion Heart) in 1199. And the legend of Robin Hood was purportedly based in part on the exploits of Sir Adam de Gurdon of Shropshire, who led a party of outlaws that robbed nobles traveling on the paths through the Hampshire forests. Some later Gurdons, including several named John, plied more respectable professions and became members of Parliament; some were even knighted.

And one uncle was, to Gurdon's good fortune, a fellow of Christ Church, one of the 38 colleges that make up Oxford University. Gurdon took the entrance exam for Latin and Greek at Oxford, and was told that he could enter the college—as long as he did not study either of these subjects. His uncle intervened and an arrangement was made that allowed Gurdon to enroll at Oxford after a year of remedial private tutoring in biology.

Gurdon made it through Oxford, where he studied zoology and continued his interest in insects. As his undergraduate days wound down, he thought about pursuing a PhD in entomology. Although that did not pan out, he was invited to study developmental biology by Professor Michael Fischberg.

FINDING A MYSTERY

His first task was to find something to study.

Biologists had long appreciated the development of an animal from egg to adult as one of the most spectacular phenomena in the biosphere. The formation of a complex creature with a head, tail, limbs, brain, gut, and many other organs from a single fertilized egg cell was also one of the greatest mysteries in biology. Very little was known about the mechanisms that shaped embryonic development.

Development begins from undifferentiated cells in very early embryos and progresses to produce many kinds of differentiated cells in the adults. One of the open questions, then, was the extent to which development was a one-way process, from undifferentiated cells to differentiated cells. That appeared to be the case based on experiments that had been done decades earlier by embryologist Hans Spemann (Nobel Prize 1935). Spemann had found that when cells of an eight-cell newt embryo were separated, each could generate an entire tadpole. When parts of the embryo were separated later in development, however, they could not generate an intact animal. So it appeared that once cells differentiated they lost the ability to generate a whole animal.

Two later pioneers, Robert Briggs and Thomas King, had taken Spemann's experiment further by testing the ability of various leopard frog cell nuclei to generate a tadpole when transferred into an egg whose nucleus had been removed (or "enucleated"). They discovered that transferring the nucleus of a cell from an early leopard frog embryo to an enucleated egg could generate a swimming tadpole. They also found that transferring nuclei from germ cells—the cells that will give rise to sperm and eggs in mature frogs—from tadpoles to an enucleated egg could also give rise to healthy tadpoles. However, transferring the nucleus of a more differentiated *somatic* cell (from the body) of an older embryo to an enucleated egg produced only abnormal embryos and did not generate a tadpole. These observations raised the question of why somatic cells lost the potential to generate an entire animal.

Since the difference in cellular potential was manifested by cell nuclei, it was clear that the answer must somehow involve genes. Indeed, it was appreciated that since every species produced faithful copies of itself, and genes determined traits, it was certain that genes must guide development. But 1956, when Gurdon began his graduate work, was only three years after the discovery of the DNA double helix—years before the discovery of messenger RNA, and long before any genes would be isolated from any organism. Nothing was known about how genes actually worked.

The reason for the different potentials of early and older cells was a mystery. And that mystery, Fischberg suggested to Gurdon, was worth looking into further.

A DIFFERENT FROG, A DIFFERENT SCIENTIST, A DIFFERENT RESULT

One possibility that Fischberg, Gurdon, and other biologists considered was that somatic cells might lose genes as they differentiate. Another was that they might silence genes in some irreversible way. Gurdon wanted to find out: Do all cell types in the body have the same genes? But how could he possibly figure that out without any direct way to study genes?

Fischberg encouraged Gurdon to repeat the nuclear transfer–type experiments that had been done in the leopard frog, but with another species. He reasoned that if Gurdon obtained the same result, then the restricted potential of somatic cells would be confirmed and he could go on to ask how that happened. On the other hand, if Gurdon obtained a different result, then the question would be reopened.

The use of frogs was favored because amphibian embryos are typically large and easy to manipulate, obtainable in great quantities, and develop outside the body. For his experimental species, Gurdon chose the African frog *Xenopus laevis* primarily because, unlike the leopard frog, eggs could be obtained year-round and individuals matured more rapidly (in one year versus more than three years).

There were many technical hurdles to overcome. Gurdon had to learn how to remove the egg cell nucleus, and how to pick up single donor cells and microinject their nuclei into the eggs. He also needed to be sure that no egg chromosomes remained that might contribute to development, and to be able to verify that embryos that developed after nuclear transfer of the donor cell were derived from the donor nucleus.

Two discoveries in the Fischberg laboratory proved invaluable. First, the lab acquired a microscope with an ultraviolet (UV) light source. Gurdon hoped,

1-Nucleolated Strain 2-Nucleolated Strain

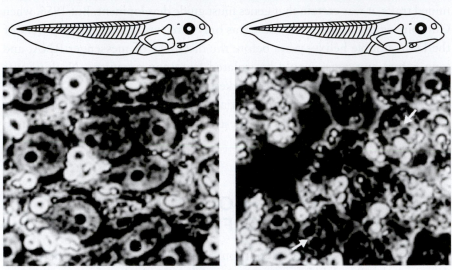

FIGURE 17.1 A Nucleolar Marker in *Xenopus laevis* The one-nucleolated strain has
only one nucleolus per diploid nucleus (left), compared with the wild-type two-nucleolated
form in which most nuclei have two nucleoli (right; arrows).

and was able to verify, that aiming the UV light on the egg nucleus destroyed
the egg chromosomes. And, second, a fellow student Sheila Smith obtained a
mutant frog that contained only one nucleolus per diploid cell, instead of the
normal two (the nucleolus is a structure in the nucleus that contains the genes
for making part of the ribosome, but this was not known at the time). Fischberg realized that the single nucleolus could be used as a visible marker in
transplant experiments to distinguish donor (one nucleolus) from recipient
(two nucleoli) cells (Figure 17.1).

Gurdon conducted the nuclear transfer experiments using the same spectrum of donor cells that had been used by Briggs and King. Like Briggs and
King, he found that he could reliably produce tadpoles with donor nuclei taken
from early *Xenopus* embryos at the blastula or gastrula stage, about 36% of
transfers (Figure 17.2). But in contrast to Briggs and King, who never obtained
normal tadpoles with older intestinal cell nuclei, Gurdon was able to obtain
10 normal tadpoles in more than 700 attempts.

That was just 1.5% of transfers but that small fraction was a big difference
from none at all. Gurdon's results demonstrated that somatic cells could be

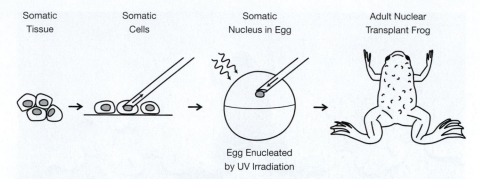

Somatic Tissue   Somatic Cells   Somatic Nucleus in Egg   Adult Nuclear Transplant Frog

Egg Enucleated by UV Irradiation

FIGURE 17.2 Somatic Cell Nuclear Transfer A somatic cell such as an intestinal epithelial cell is isolated in a micropipette and injected into an unfertilized frog egg whose nucleus has been removed by ultraviolet irradiation.

Adapted with permission from Gurdon (2013). Illustration by Kate Baldwin.

"reprogrammed" by factors present in the egg to generate all of the tissues and cells of the body. The normal tadpoles indicated that differentiation was not irreversible and that cells did not lose genes as they differentiated.

There was skepticism that a graduate student working alone could repeat, let alone overturn, the pioneering work conducted by Briggs and King. One key difference between Gurdon's efforts and his predecessors was that he attempted many more transfers of older nuclei than his predecessors, and thus was able to observe and repeat the rare success.

SEND IN THE CLONES!

Following his graduate work, Gurdon spent some time in the United States. While he was away, his former advisor Fischberg looked after his frogs, which matured into fertile males and females—all genetically identical to, and thus *clones* of, the donor frog. This was the first demonstration that fertile adult animals could be obtained by the reprogramming of their somatic cells. Gurdon later used an albino frog as a donor, and created a flock of cloned frogs for a stunning photograph (Figure 17.3).

In addition to science fiction story lines, there were potential practical implications to using somatic cell nuclear transfer, for example, to regenerate body tissues. However, it would be more than 30 years before the cloning of a mammal from a somatic cell was achieved. This was due primarily to the much smaller size of mammalian eggs (about 1/1000 the volume of an amphibian egg), which were very difficult to manipulate. The first success was obtained

FIGURE 17.3 Cloned Frogs Here, clones of an albino male frog were obtained by transplanting nuclei from cells of an albino embryo into enucleated eggs of the wild-type female shown. The albino frogs are genetically identical.

From Gurdon (2013). Republished with permission of Company of Biologists, Ltd. © 2013; permission conveyed through Copyright Clearance Center, Inc.

in 1996 when mammary gland nuclei were used to produce a female sheep named Dolly. Since that time, adult donor cell nuclei have been used to clone mice, cows, goats, pigs, rabbits, and cats, among others.

These successes demonstrated that the ability of mature cell nuclei to be reprogrammed when placed in an enucleated egg cell was widely shared among species. It was not known, however, whether a mature cell could be reprogrammed without the presence of the egg cell.

In 2006, Shinya Yamanaka and co-workers reported the stunning discovery that mouse fibroblasts (cells that make connective tissue) could regain the ability to produce all cell types in the body by the addition of just four proteins, called *transcription factors*, that regulate gene expression. Yamanaka's team also demonstrated that the same four factors could reprogram human fibroblasts. He called these reprogrammed cells "induced pluripotent stem cells" (iPSCs; pluripotent means that they have the potential to produce most

FIGURE 17.4 Shinya Yamanaka and John Gurdon, 2012 AP Photo/Bertil Enevag Ericson.

or all cell types). These discoveries paved the way to generating virtually any cell type from an individual, and the use of iPSCs is being investigated in many clinical applications.

In 2012, Yamanaka, who was born the same year that Gurdon published his breakthrough paper (1962), shared the Nobel Prize in Physiology or Medicine with Gurdon for their discoveries "that mature cells can be reprogrammed to become pluripotent" (Figure 17.4).

END-OF-CHAPTER QUESTIONS

1. Why did some researchers think that cells might lose or irreversibly silence genes in the course of development?
2. What did Gurdon's advisor think could be gained by repeating the experiments done previously by Briggs and King?

3. Why was Gurdon successful in obtaining normal tadpoles and frogs from somatic cell nuclear transfer? How could he prove that the frogs were of donor nuclei origin?

4. Your beloved dog Spot is getting on in years and you can't bear the thought of losing her. Design a procedure that would enable you to create a new, genetically identical Spot.

18

THE ARSENAL OF IMMUNITY

To create is to recombine.

—FRANÇOIS JACOB, NOBEL LAUREATE 1965

EVEN BEFORE BIRTH, we are exposed to bacteria in the placenta. We pick up more microbial hitchhikers during birth, and from then on, we live in a world populated by millions of species of bacteria, fungi, and viruses.

By the time we become adults, there are more bacterial cells in and on our bodies (about 40 trillion) than our own. They are members of about 1000 or so species that, along with species from about 80 genera of fungi, make up the human *microbiome*. And those are just the typical residents. Just think about all of the coughing and sneezing that goes on.

How do we defend ourselves against the hordes of foreign invaders?

The answer is our immune system. We and all vertebrates—fish, amphibians, reptiles, birds, and mammals—possess an "adaptive" immune system. It is called adaptive because it is able to respond to virtually any foreign invader or substance by producing *antibodies* specific to that foreign agent. This is the foundation of the long-established practice of vaccination. But the source of that power—how the immune system is able to manufacture potentially millions of different antibodies—was a long-standing mystery that captivated and eluded large numbers of researchers, among them many of the most accomplished immunologists in the world.

Thirty-five-year-old Susumu Tonegawa was not among the latter. Indeed, he had no formal training in immunology whatsoever. Yet it was Tonegawa who would make the stunning discovery that solved the mystery of antibody diversity.

Breakthroughs in science happen in many different ways. Sometimes, it is a new kind of experiment—transplanting nuclei or hurling starfish—that reveals a new phenomenon. Sometimes, the invention of new technology such as microscopes or telescopes allows us to see what was previously invisible. Tonegawa was propelled by both new experiments and powerful new technology. In fact, Tonegawa took advantage of no fewer than four contemporaneous Nobel Prize–winning advances in making his breakthrough.

THE PARADOX

The importance and the mystery of the immune system had drawn great medical and scientific interest since the 19th century. By the 1970s, it was well established that the body's immune system could mount a response to any invader—bacteria, viruses, fungi, or parasites—or to almost any foreign substance such as proteins or complex carbohydrates. It was also well known that one major branch of the immune response involves a class of white blood cells called B cells. Once activated by a foreign agent or *antigen*, B cells differentiate into plasma cells that secrete antibody proteins that bind specifically to the antigen. Once bound, antibodies help to kill or clear foreign agents from the body. In rare cases of people who are unable to make antibodies, they suffer from severe, recurrent, and life-threatening infections.

The structure of antibody proteins was also well understood. They are Y-shaped molecules composed of four polypeptide chains: two longer "heavy" and two shorter "light" chains (Figure 18.1). The Y shape emerges from how the chains are put together. The heavy and light chains bind to one another and are held together by disulfide bonds, and the heavy chains themselves are also held together by disulfide bonds. The amino-terminal ends of the heavy and light chains combine together to form a three-dimensional pocket called the antigen-binding site, of which there are two per individual antibody molecule. The specificity of antibodies is determined by the sequence of amino acids in the antigen-binding site.

What was not at all clear was how the body was able to produce such a diverse array of antibodies specific for virtually any antigen. In fact, it had been shown that antibodies could even be produced that were specific for synthetic chemicals not present in nature. The potential chemical diversity of antigens a person or animal might encounter is so enormous, the immune system would require millions (perhaps tens or hundreds of millions) of different antibodies to be able to recognize them all.

That diversity presented a genetic paradox. Antibodies are proteins so they are encoded by genes. Did the genome really contain millions of antibody genes, that is, millions of heavy and light chain genes? That seemed difficult to imagine because the genetic code for 1 million heavy and light chain genes would occupy at least 2 billion base pairs of DNA. That would be at least two-thirds of the entire human genome. On the other hand, it was also difficult to understand what other mechanism could generate such diversity.

Two major hypotheses had been put forward. The *germ line theory* proposed that the genes for every antibody molecule were encoded in the genome, and that there was a very large number of heavy and light chain genes. The *somatic theory* suggested that some events took place in the development of B cells that generated antibody diversity from a more modest number of heavy and light chain genes. However, there was no precedent for such a mechanism, and any rearrangement of genes in somatic cells would violate the generally accepted principle Gurdon had established that somatic cells have the same DNA content as the germ line (Chapter 17).

INTO THE GREAT WIDE OPEN

The challenge facing Tonegawa when he first arrived in 1971 at the Basel Institute for Immunology in Switzerland was that there were no tools to get to antibody genes. No eukaryotic gene had been isolated. And there were no techniques available that would allow the determination of antibody gene number or structure in animals. He would have to improvise and innovate.

One great source of insights into antibody structure and function had come from immunologists' study of a particular kind of plasma cell cancer called a myeloma that produces antibodies. Studies had shown that the antibody molecules made by a given myeloma were all identical. However, examination of the amino acid sequences of many different myeloma antibody molecules revealed that the amino-terminal regions of both the light and heavy chains varied between antibodies, while the remainder of each protein chain showed little variability. These two regions of the antibody chains were dubbed the variable (V) and constant (C) regions (V_L and C_L, and V_H and C_H; see Figure 18.1).

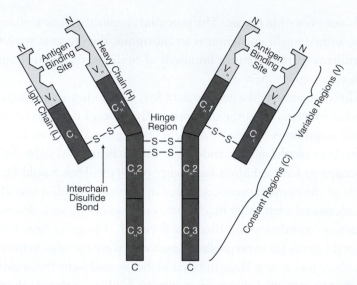

FIGURE 18.1 The Structure of Antibodies Each antibody consists of four polypeptides: two heavy (H) chains and two light (L) chains joined to form a Y-shaped molecule. The chains are held together by disulfide bonds (–S—S–). The amino-terminal regions of each chain form the antigen-binding site. These two regions of each chain are more variable in sequence and are dubbed the variable (V) regions; the rest of each chain is called the constant region (noted as V_L and C_L; V_H and C_H). Illustration by Kate Baldwin.

Everyone realized this was an important clue to the generation of antibody diversity. Whatever the explanation might be, it had to account for how one end of the protein was much more variable than the other. However, the existence of the variable and constant regions did not alone favor either the germ line or somatic theories for antibody diversity.

The key issue remained the number of light and heavy chain genes. Just the ballpark number was important: were there millions or a much smaller number?

Tonegawa came up with an idea to try. A method for purifying a specific messenger RNA (mRNA) was just coming into practice. And techniques for estimating the relative number of copies of a DNA sequence based on the kinetics of DNA hybridization had been pioneered. Tonegawa purified a light chain mRNA from a myeloma cell line, labeled it with radioactivity, and studied its hybridization kinetics with whole mouse genomic DNA. He also studied an mRNA for a gene that was known to be present in just one or two copies. His results indicated that the constant regions of the light chains were encoded by relatively few genes. This result suggested to Tonegawa that since the germ line does not contain many light chain genes, there must be some somatic mechanism operating.

A PLEXIGLASS TRAY

A rough count of light chain gene number in whole genomic DNA was a long way from understanding any mechanism at the molecular level. Tonegawa needed some way to hone in on the antibody genes themselves. He tried to imagine some way to separate the antibody genes from the bulk of genomic DNA.

One timely advance was the discovery of restriction enzymes in bacteria that cleave double-stranded DNA at specific sequences—a discovery that would help launch the era of genetic engineering and earn the 1978 Nobel Prize in Physiology or Medicine. Tonegawa realized that he might be able to cleave mouse DNA into much smaller fragments that could contain single antibody genes. Detecting those genes, however, would require some way of separating the thousands of fragments of DNA from one another that had not yet been invented.

One day, he happened to spot the potential solution in a refrigerated room at the institute. Someone had poured starch gel into a huge plexiglass tray, and was using an electric field to separate serum proteins based on their size and charge (electrophoresis). Tonegawa thought that the same technique might be applicable to separating DNA fragments. If so, he could then go searching among the fragments for those that encoded antibody genes. Most importantly, he imagined that if some rearrangement took place in antibody-producing cells, the genes encoding antibody genes might be on different DNA fragments in germ line and myeloma cells.

This was before restriction enzymes were commercially available, so Tonegawa and postdoctoral fellow Nobumichi Hozumi had to produce and purify their own enzyme. Then, they poured 2 liters of melted agarose into the plexiglass tray, allowed it to solidify, inserted a partition that divided the gel in half, and cut a well into the top of the gel on each side of the partition. Into each well they placed 5 milligrams of restriction enzyme–digested embryonic DNA or mouse myeloma DNA, and electrophoresed the DNA for 3 days. Afterward, they cut each side of the gel into 30 thin slices, extracted the DNA, and hybridized it with radioactively labeled light chain RNA or its 3′ fragment (Figure 18.2).

The result was as clear as they could have dreamed. The light chain mRNA hybridized to DNA from two slices of embryonic DNA, but to just one entirely

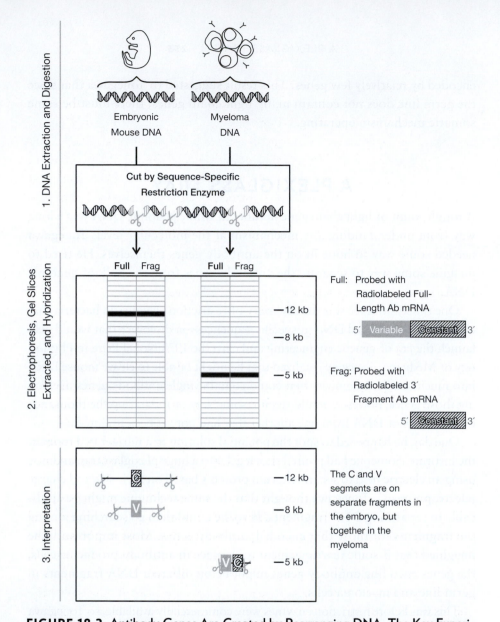

1. DNA Extraction and Digestion

Embryonic
Mouse DNA

Myeloma
DNA

Cut by Sequence-Specific
Restriction Enzyme

2. Electrophoresis, Gel Slices Extracted, and Hybridization

Full   Frag   Full   Frag

—12 kb

—8 kb

—5 kb

Full: Probed with
        Radiolabeled Full-
        Length Ab mRNA

5′ Variable Constant 3′

Frag: Probed with
      Radiolabeled 3′
      Fragment Ab mRNA

5′ Constant 3′

3. Interpretation

—12 kb

—8 kb

—5 kb

The C and V
segments are on
separate fragments in
the embryo, but
together in the
myeloma

FIGURE 18.2 Antibody Genes Are Created by Rearranging DNA: The Key Experimental Evidence Hozumi and Tonegawa discovered that the variable (V) and constant (C) regions of antibody genes were arranged differently in embryonic DNA and myeloma cell DNA. The critical experiment involved several steps: (1) The DNA was extracted from embryonic tissue or cells and digested with a restriction enzyme that cut the DNA at specific 6–base pair sequences. (2) Fragments of the digested DNA were separated by electrophoresis in an agarose gel. The gel was then sliced into 30 slices, and the DNA was extracted and hybridized with radioactively labeled probes that were either the full-length (Full) antibody mRNA or a fragment (Frag) that included just the constant region of the mRNA. (3) The segments of DNA that contained the C and V regions were in different gel slices in embryonic DNA but in the same gel slice in myeloma DNA, indicating that the C and V gene segments rearranged during the development of antibody-producing cells. Illustration by Kate Baldwin.

different slice from myeloma DNA. This revealed that the V and C regions are on separate DNA fragments in the embryo, but are located on the same DNA fragment in the myeloma (Figure 18.2). Thus, the V and C regions of the light chain gene were closer together in antibody-producing cells.

The result stunned fellow immunologists and biologists in general. Tonegawa was a relative unknown, yet he had obtained powerful evidence that antibody genes rearranged and recombined during development. It was also the first evidence of any such process in animals. This scoop was just a first act.

SEEING IS BELIEVING

The plexiglass tray experiment gave an elegant result, but it was still a very low-resolution view of the dynamics of antibody genes. To understand the rearrangement process, it was necessary to study individual embryonic and myeloma light or heavy chain genes and their sequences. Those techniques also came along just in time. DNA cloning, or the isolation of individual DNA fragments in microbial hosts, was initially a controversial subject when invented in 1972 (another Nobel Prize–winning accomplishment). After biologists took a pause to assess the safety and ethical issues, DNA cloning resumed under a tightly regulated set of rules.

Tonegawa and his team first cloned DNA containing light chain genes from embryonic and myeloma cells. He thought that one fairly fast and powerful way to compare the arrangement of genes within the clones was to see how each hybridized to a light chain mRNA. Seeing in this case was literal, because RNA–DNA hybrids can be visualized in the electron microscope.

Tonegawa himself had no expertise in the technique so he went to see Christine Brack, a highly accomplished electron microscopist at another institute in Basel, and asked if she would like to join the effort. Brack promptly agreed, and within just a few weeks of starting her work, she came to Tonegawa with a beaming smile and showed him very clear images of molecules that she had photographed (Figure 18.3). Under the right conditions, the RNA hybridizes with its complementary strand of DNA hybrid. Wherever along the DNA that the RNA does not have a complementary sequence to pair with, the single-stranded DNA forms loops visible in the electron microscope.

Those loops clearly revealed that V and C regions are indeed separate in embryonic DNA and are brought close together in the myeloma cell. But to Tonegawa's and everyone's surprise, they were not brought immediately

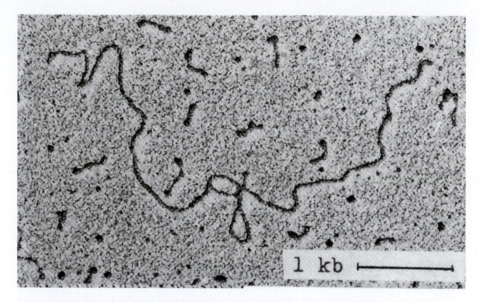

FIGURE 18.3 RNA–DNA Hybrids Reveal Segments of Antibody Genes Electron micrograph of light chain mRNA hybridized to cloned light chain DNA. The linear, string-like shape is the RNA–DNA hybrid. Wherever along the DNA that the RNA does not have a complementary sequence to pair with, it forms visible loops.

Reprinted from Brack et al. (1978). © 1978 with permission of Elsevier.

adjacent to one another. There was a DNA region between the V and C regions that was not present in the mRNA. This was the first example in animals of an intron, a segment of DNA within a gene that is not represented in the mature mRNA transcript. The first examples had been found in viruses just a few months earlier (the discovery would earn the 1993 Nobel Prize in Physiology or Medicine).

To understand the precise structure of the genes—which bases encoded which part of the protein—would require determining their exact sequence. No method had yet been published for sequencing DNA; however, two laboratories had developed two different ways of doing so (these pioneers would promptly win the 1980 Nobel Prize in Chemistry). Tonegawa teamed up with one group to determine the first sequences of antibody genes.

DNA sequencing revealed another surprise. The region of DNA encoding a short sequence of the light chain at the end of the V region was located in its own region *between* the V and C segments. The short segment was dubbed the J region for "joining" because the joining of the V and C segments occurred right at the J segment. Thus, light chains were put together not from two separate pieces of DNA, but from three—the V, J, and C segments.

This detailed, segmental structure of an individual light chain gene revealed that an antibody chain was assembled from pieces. The key questions for understanding antibody diversity, however, still remained: Just how many of these pieces were there? And, how many different antibodies could they make?

ASSEMBLING THE ARSENAL: THE POWER OF COMBINATIONS

With the technology of DNA cloning and analysis spreading quickly, knowledge of antibody genes grew quickly, too. Analysis of heavy chain genes, which are located on a different chromosome than light chain genes, revealed that individual chains are assembled from four gene segments. In addition to V, J, and C segments, a fourth D segment is included between the V and J segments (Figure 18.4).

The body's total arsenal of antibody genes is a straightforward calculation, similar to calculating the number of possible hands in a card game. Given the number of cards in each suit, and the number of suits, one can calculate how

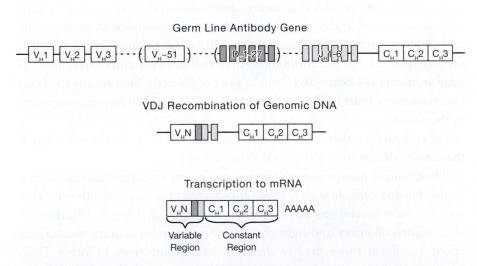

FIGURE 18.4 Combinatorial Assembly of Antibody Genes from Gene Segments Schematic of the arrangement of heavy chain gene segments in the human germ line. During B cell development, one of ~51 V segments is joined with one of 27 D segments and one of 6 J segments, and with a constant region, to assemble a complete heavy chain gene. Antibody diversity is a product of the large number of random combinations of gene segments available for heavy and light chain genes. Illustration by Kate Baldwin.

many different three-, four-, or five-card hands one could draw at random. The same follows for antibody genes. By knowing how many V, J, or D gene segments there are, and assuming they are recombined at random to make individual V regions, one can calculate the number of possible heavy and light chains. (The C regions are important for other antibody functions but do not contribute to antigen binding, so they are not part of the diversity calculation.)

For example, it turns out that humans have 51 heavy chain V segments, 27 heavy chain D segments, and 6 heavy chain J segments. Therefore, if each of these is used and combined at random to assemble V regions of heavy chains, humans can make $51 \times 27 \times 6 = 8262$ different heavy chain V regions. That is a lot of heavy chains from just 84 $(51 + 27 + 6)$ gene segments.

Humans also have 40 V segments and 5 J segments for one type of light chain (called kappa), and 30 V segments and 4 J segments for a second type of light chain (called lambda). Again, if these are assembled at random, then there are $40 \times 5 = 200$ different kappa V regions, and $30 \times 4 = 120$ different lambda V regions. That makes a total of 320 light chain V regions from 79 gene segments.

One more very important calculation gets us to the total number of possible antibodies. It is the combination of individual light chain and individual heavy chain polypeptides that creates antibodies and their antigen-binding sites. Assuming light chains and heavy chains are combined at random, and all combinations are used, then there are 320×8262, or more than 2.6 million different possible human antibodies. All of that diversity comes from just 163 gene segments (84 heavy and 79 light gene segments). That means the body can make more than 10,000 times as many antibodies as it has gene segments in the genome.

For comparison, there are almost 2.6 million possible five-card poker hands that can be drawn from a standard 52-card deck.

Thus, simple math shows the impressive power of combinatorial mechanisms. Plasma cells, however, have supercharged the diversification process with one more special genetic trick. Within activated, dividing B cells, the variable regions of heavy and light chains undergo further mutations at a rate about 1 million times greater than the background rate of other DNA sequences. This "somatic hypermutation" process expands the arsenal of antibody diversity by at least another 10-fold.

Physicist and Nobel laureate Jean Perrin said that the key to scientific advances is to be able to "explain the complex visible by some simple invisible." Propelled by ingenuity and operating right at the frontier of what was experimentally possible, Tonegawa allowed us to see for the first time how from a small number of invisible genes, the body assembles a powerful arsenal capable of combating anything it encounters. For his discovery of the genetic principle underlying the generation of antibody diversity, Tonegawa was awarded the 1987 Nobel Prize in Physiology or Medicine.

END-OF-CHAPTER QUESTIONS

1. Four independent Nobel prize–winning discoveries helped Tonegawa solve the mystery of antibody diversity. What were they? In what way were they essential to advancing the understanding of antibody diversity?
2. A lottery is conducted by drawing a series of balls numbered 1–20 at random from a tumbling basket. How many different sequences of numbers can be generated by drawing three balls? Four balls? Six balls? Discuss the power of combinations in the context of the immune system.
3. Many years after Tonegawa's breakthrough, two enzymes were identified that are responsible for catalyzing the VDJ recombination process. Predict the medical problems that would result from lacking either of these two enzymes.

19

EVERYONE HAS A SPLIT PERSONALITY

*The great pleasure and feeling in my right brain is more than
my left brain can find the words to tell you.*
—ROGER SPERRY, NOBEL LAUREATE 1981

BEFORE THERE WERE SEARCH engines, before Kindles, before Wikipedia, before iPods, and long before Google Maps, there was Kim Peek.

By the time Peek died in 2009 at age 58, he knew 12,000 books by heart, and had encyclopedic knowledge of at least 15 subjects, including world and American history, geography, space exploration, literature, sports, Shakespeare, and the Bible. He could identify hundreds of works of classical music, knew every zip code in the United States, and could provide travel directions between any pair of major cities. Given any past date, he could say almost instantly the day of the week on which it fell.

Peek amassed such a vast reservoir of knowledge by consuming information like no other human being. He started at 18 months old, memorizing books as they were read to him. Later, with the ability to read each page of an open book simultaneously, one page with his left eye, the other with his right—in just eight to ten seconds—he could down a 500-page novel in a little more than an hour. Even at that speed, Peek would have almost total recall of characters, even quotes, months and years later. He would replace a book upside down on his bookshelf once it was "recorded" (Figure 19.1A).

Yet, when Peek's IQ was tested at age 27, his overall score was 87, below the average of 100, although certain subtests were in the superior range. Peek

(A)

Normal Brain

Kim Peek's Brain

Corpus
Callosum

Anterior
Commissure

Posterior
Commissure

Cerebellum

(B)

FIGURE 19.1 Kim Peek and His Exceptional Brain (A) Kim Peek. (B) MRI images of a typical brain and Kim Peek's brain. Kim Peek's brain differs from typical brains in that the cerebellum is smaller than usual and malformed. Most striking is the absence of the largest commissure, called the corpus callosum; the anterior and posterior commissures are also missing. (A) ITV/Shutterstock; (B) photo courtesy of Pratik Mukherjee, MD, PhD.

struggled with motor skills and coordination, he shuffled when he walked, and had difficulty buttoning his clothes and performing most routine daily chores. Peek was thus described as a *savant*—a person with a developmental disorder who exhibits extraordinary ability in certain fields. It is estimated that perhaps 10% of people on the autistic spectrum exhibit some savant skills—in memory, mathematics, or calendar computing—but very, very few have the prodigious powers Peek had.

The functioning of just the typical human brain with our unique capability in the animal kingdom to produce language and speech, to create art and music, and to reason poses some of the most fascinating and still unsolved mysteries in biology. Peek's extraordinary abilities, and how they could coexist with severe disabilities, are yet a deeper mystery.

Most advances in biology have come from experiments in model species, and from the replication of experiments—transforming bacteria, injecting frog eggs or dog retinas, etc. The human brain, however, has unique capabilities, so how can these be approached if not represented in other species? While common aspects of brain function have been tackled through animal studies, some specific powers of the human brain have been illuminated by studies of a small number of unique people like Peek with unusual abilities or conditions.

MISSING A CONNECTION

Naturally, scientists and doctors were very interested in Peek's remarkable brain, and sought some anatomical explanation for his abilities. A magnetic resonance imaging (MRI) scan revealed that his brain was large, in the 99th percentile, but several areas were not normal. The cerebellum, the part of the brain involved in balance, was malformed, which probably accounted for some deficits in coordination. Most striking was the absence of the corpus callosum, a bundle of nerve tissue that connects the right and left hemispheres of the brain (Figure 19.1B).

This structure has figured prominently in early functional studies of the human brain and in medicine. In 1939, surgeon William Van Wagenen of the University of Rochester Medical Center sought to treat ten patients with severe epilepsy who suffered from violent and uncontrollable seizures. Van Wagenen had previously observed seizures lessen in two brain tumor patients when the tumor destroyed the corpus callosum. He reasoned that since the waves of debilitating brain activity swept across both brain hemispheres, and the roughly 250 million nerves of the corpus callosum conducted most of the electrical traffic between the left and right hemispheres, severing that connection might help the epilepsy patients.

There was no precedent for such a radical operation; however, Van Wagenen and the patients were desperate. He accessed the corpus callosum from the top of the skull, a good portion of which had to be opened to reach the length of the cable-like structure deep within the brain. The surgery did in fact help most of the patients by reducing the frequency and severity of seizures. Amazingly, psychological testing revealed no major changes in intelligence, memory, motor skills, or behavior. Despite their severed corpora callosa, and thus disconnected brain hemispheres, the patients were reported to be "normal."

How could the largest connection and one of the most prominent features of the brain apparently matter so little to its normal function? Since people appeared to be able to function fine without it, some observers at the time quipped that the corpus callosum only seemed to serve to propagate epileptic seizures, or perhaps just to hold the two hemispheres together.

The paradox attracted the interest of neurobiologist Roger Sperry in the early 1950s. Sperry found it difficult to accept that the corpus callosum appeared dispensable. He and his students launched a series of studies into its function in animals by surgically severing the bundle of fibers and the connections between the left and right hemispheres, first in cats and then in primates.

The so-called split-brain animals exhibited little or no change in ordinary behavior, even when other smaller connections between the hemispheres were also cut. The animals were as alert and active, were as coordinated, and maintained the same temperament as untreated animals.

Sperry and his team learned, however, that looks were deceiving, at least with respect to how the split brain functioned. When they designed tests that were able to distinguish the activities of each brain hemisphere, they discovered that what was learned by one side of the brain was not transferred to the other side.

For example, a cat had both its corpus callosum and optic chiasm cut (the chiasm is another connection through which the optic nerves cross over from each eye to the opposite brain hemisphere). Then, it was trained to respond to a geometric shape such as a circle or square by pushing a panel to obtain a food reward. The twist, however, was that the shape was presented to only one eye. Once the cat had mastered the training, the first eye was blindfolded and the same shape was presented to the other eye. The cat behaved as if it had never seen the shape.

The same was true for other brain systems such as those based on touch. Sperry and a colleague trained cats to get food by choosing one of two foot

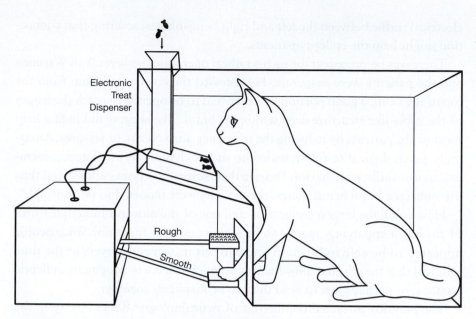

FIGURE 19.2 Testing Cognition in Split-Brain Cats Sperry and colleagues designed a test in which a cat learned to distinguish between two pedals with different surfaces using its paw to earn a food reward. Cats in which the corpus callosum was severed could not transfer the learning to the other paw. Adapted from Sperry (1964). Illustration by Kate Baldwin.

pedals they could not see—a hard one or a soft one—using their forepaws (Figure 19.2). Cats with an intact corpus callosum that learned to use one paw were able to do the task when forced to use the other paw. But in cats with a severed corpus callosum, the task had to be learned anew.

The response to a given test depended on which side of the brain was stimulated. "It was as though each hemisphere were a separate mental domain operating with complete disregard—indeed, with a complete lack of awareness—of what went on in the other," Sperry wrote.

Tests with monkeys produced similar results, indicating that the corpus callosum was generally required in mammals for enabling the two hemispheres to share learning and memory. Naturally, Sperry wanted to know whether the same was true of humans. So did Michael Gazzaniga, a Dartmouth undergraduate who spent the summer before his senior year (1960) in Sperry's lab at Caltech.

Van Wagenen's split-brain patients had exhibited no obvious deficits. But had Van Wagenen looked hard enough, with the right tests? Gazzaniga wanted to retest the Rochester patients, who were apparently the only people to have had the split-brain surgery. Even though the surgery had been generally successful, and the patients did well afterward, for some reason the surgery had not been repeated elsewhere.

Gazzaniga tracked down one of the attending physicians from 20 years earlier, who agreed to give him the patients' information so they might be located. He and Sperry designed some experiments. Then, Gazzaniga headed to Rochester with a carful of psychological testing equipment, only to be told by the physician once he got there that he had changed his mind and would not grant Gazzaniga access to the records.

Although Gazzaniga returned from Rochester empty-handed, he remained determined to study the brain. He finished his undergraduate degree, applied to graduate school at Caltech, and joined Sperry's lab full-time.

WILLIAM JENKINS

Gazzaniga received his assignment on the first day of graduate school: to study split-brain patients at Caltech. The only hitch was that there were no split-brain patients anywhere except for the Rochester group to which he had been denied.

That problem was soon solved. In the summer of 1960, neurosurgery resident Dr. Joseph Bogen was on call at a Los Angeles hospital when a 48-year-old male patient was brought to the emergency room with a very severe seizure. Bogen learned that William Jenkins was a World War II veteran who had parachuted behind enemy lines, been captured, and struck with a rifle butt. The seizures had become progressively worse in the 15 years since the injury, as he suffered as many as 7–10 episodes a day. No medications had shown any benefit.

Over the next several months, Bogen took a keen interest in Jenkins' case. Bogen had previously spent a postdoctoral stint at Caltech, and his advisor had an office next to Sperry's. Bogen knew about Sperry and Van Wagenen's split-brain surgeries and thought that the procedure might help Jenkins. The former paratrooper was willing to undergo the operation. He told Bogen, "You know, even if it doesn't help my seizures, if you learn something it will be more worthwhile than anything I've been able to do for years."

Bogen persuaded the chief of neurosurgery to consider performing the procedure. They decided it would be wise to practice the operation first—in the morgue. Bogen met Gazzaniga during this time. They became fast friends and designed a series of tests to be conducted before and after the surgery.

Performed in February 1962, the operation was very successful in curtailing Jenkins' seizures. Moreover, Jenkins did not appear to suffer any changes to his intellectual abilities or personality. In fact, Jenkins said that he felt better than he had in many years. Sperry noted that in casual conversation over a cup of coffee "one would hardly suspect that there was anything at all unusual about him." It would be up to Gazzaniga to find out.

ONE BRAIN, TWO MINDS

One afternoon, Gazzaniga began a series of tests of hemisphere function. He started with visual processing. Gazzaniga asked Jenkins to look at a small dot of paper attached to a screen. Then, to the right of the dot, he flashed a picture of a square for just 100 milliseconds. It was known that the visual field of each eye was processed by the opposite brain hemisphere. The square would be processed by the left hemisphere.

"What did you see?" Gazzaniga asked.

"A box," replied Jenkins.

"Good, let's do it again."

Gazzaniga flashed another image of a square, but this time to the left of the dot, where it would be processed by the right hemisphere.

"What did you see?" Gazzaniga asked.

"Nothing," Jenkins replied.

"Nothing. You saw nothing?" Gazzaniga pressed.

"Nothing," Jenkins repeated.

Gazzaniga's heart raced. It was the moment of discovery. Gazzaniga had to process what he had just witnessed. With the connections between the two hemispheres severed, the right hemisphere could not communicate what it saw to the left hemisphere, which controls speech (Figure 19.3). It was as if there were two minds operating in one head, one that could speak, and one that could not.

Gazzaniga tried a different test, this time flashing a circle to either side of the dot. He asked Jenkins to point to what he saw using whichever hand he wanted. In this test, Jenkins was able to point to the shape on each side. He used his left hand to point to the circle flashed on the left, and his right hand to point to an image flashed on the right. This meant the right hemisphere could see the image and mount a nonverbal response with the hand whose movement

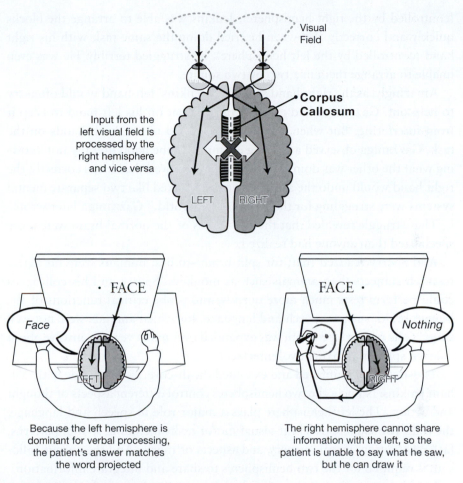

FIGURE 19.3 Testing Cognition in Split-Brain Patients

Adapted from Wolman (2012). Reprinted by permission from Macmillan Publishers Ltd: Nature. Copyright © 2012, Springer Nature. Illustration by Kate Baldwin.

it controlled (the right hemisphere controls the left hand; the left hemisphere controls the right hand). But the right hemisphere could not say what it saw.

Gazzaniga later extended the tests to the sense of touch. Jenkins was blindfolded so that he could not see what objects Gazzaniga was placing into his hands. When an object was placed into his right hand, Jenkins had no trouble identifying it. But when an object was placed in his left hand (controlled by the right hemisphere), he could not name it.

The right hemisphere, however, was perfectly able to complete other visual-motor tasks. Gazzaniga gave Jenkins four multicolored blocks, then showed him a picture of the order in which to arrange them. Using his left hand (controlled by the right hemisphere), Jenkins was able to arrange the blocks quickly and correctly. But when he tried doing the same task with his right hand (controlled by the left hemisphere), he struggled terribly. He was even unable to arrange them in a two by two square.

Amazingly, as the right hand struggled, Jenkins' left hand would often try to help out. Gazzaniga had to ask Jenkins to sit on his left hand to keep it from interfering. But when Jenkins was allowed to use both hands on the task, Gazzaniga observed a striking example of one hand literally not knowing what the other was doing. As the left hand moved the blocks correctly, the right hand would undo the arrangement. "It looked like two separate mental systems were struggling for their view of the world," Gazzaniga later wrote.

That struggle revealed that the two halves of the normal brain were more specialized than anyone had realized.

And Sperry realized that, for split-brain studies, humans were far better research subjects than animals such as monkeys. Sperry and his colleagues could perform tests much more quickly, and probe critical functions of the human brain, such as speech and language, for which animals were not well suited. The research program was expanded over many years to involve about a dozen split-brain patient volunteers.

Those studies confirmed and extended the discoveries first made with William Jenkins. Namely, the two hemispheres control different aspects of thought and action. The left hemisphere plays a major role in speech and language; the right hemisphere excels at visual-motor tasks including spatial patterns, face recognition, visual imagery, and aspects of music. And the corpus callosum is required for the two hemispheres to share and integrate information.

For his pioneering discoveries on the specialization of the brain hemispheres and insights into the "inner world of the brain," Sperry shared the 1981 Nobel Prize in Physiology or Medicine.

RAIN MAN

Still, despite Sperry and colleagues' insights, how can one explain Kim Peek, who was born without a corpus callosum and yet could record, retrieve, and communicate information from both hemispheres? Aware of such savants, Sperry anticipated what must be part of the explanation—that Peek's brain must have compensated somehow as it developed. A brain that never had a corpus callosum might be expected to be wired and to function differently than one that developed the normal hemispheric connections and then had them cut.

The source of Peek's remarkable powers is not known, but his story is. In 1984, screenwriter Barry Morrow met Peek at a gathering in Texas. Over the course of a few hours, Peek astonished Morrow with his incredible memory and lightning-fast calculations. The encounter inspired Morrow to write the movie *Rain Man*, a fictional story about an autistic savant, played by actor Dustin Hoffman. To prepare for the role, Hoffman spent a day with Peek.

The very popular 1988 film was nominated for eight Academy Awards. When Hoffman took the stage to accept his Oscar for best actor, he gave "special thanks to Kim Peek for his help in making *Rain Man* a reality." Barry Morrow went so far as to give his screenwriting Oscar statuette to Peek.

Hoffman's statement spurred enormous interest in Peek, the real-life inspiration for the movie. Hoffman encouraged Peek's father to share his remarkable son with the world, and so he did. By the time Peek passed away in 2009, he had appeared before hundreds of thousands of people around the world, demonstrating his abilities and advocating for the disabled.

END-OF-CHAPTER QUESTIONS

1. Why was Sperry skeptical of Van Wagenen's report of the normal abilities of human split-brain patients?
2. How was Michael Gazzaniga able to perceive changes in split-brain patients that Van Wagenen did not?
3. Some experts have argued that we really don't understand the brain until we can explain savants like Kim Peek. What are some possible explanations for savant abilities? How might they be tested?

APPENDIX:
THE SCIENTIFIC PROCESS

The scientific process refers to a well-established series of steps for understanding and explaining the natural world. The steps highlighted in bold below form the basis of most scientific inquiry, including every story in this book.

Observation

The scientific process usually begins with some kind of observation. This may be noting some kind of pattern in nature, such as the distribution of species on a group of islands, or the correlation of one phenomenon with another, such as a high incidence of cancer in people who smoke.

Questions

Observations may prompt questions about potential cause and effect. For example, why are there slightly different species on different nearby islands? Or, does smoking cause cancer?

Formulate a Hypothesis

A hypothesis is a tentative explanation for one or more observations. A useful hypothesis has two key properties:

i. It must be *testable*. There must be some way of gathering evidence that may either support or reject a hypothesis.

ii. It should make specific *predictions*. These predictions are often in the form of an "if, then" statement. For example, if smoking causes lung cancer, then smokers should have a significantly higher incidence of lung cancer than nonsmokers.

Test the Hypothesis

No matter how interesting or appealing, to be scientific, any explanation must be put to the test. Scientists must ask: Is the hypothesis consistent with existing knowledge? Is it consistent with new information? Do its predictions pan out to be true? To find out, evidence must be gathered and interpreted, which typically consists of the following activities:

1. *Collect data.* Data may be collected in an enormous variety of ways, from making additional observations to measuring any variable that may be relevant to the hypothesis at hand. For example, in assessing the hypothesis of smoking causing lung cancer, data could be collected about peoples' ages, gender, diets, alcohol use, the number of cigarettes smoked per day, the number of years one smoked, whether anyone else in the family smoked, any family history of cancer, and so forth.

2. *Conduct experiments.* Another way to test a hypothesis is to perform experiments. These experiments could take place in a laboratory and involve cells or live organisms, or in a natural setting, or—in the case of human testing—in a hospital. For example, one way to further test the link between smoking and cancer would be to expose cells or animals to cigarette smoke and observe or measure effects.

3. *Analyze and interpret data.* Once data are obtained, the results must be analyzed and interpreted. The key issue is whether the evidence supports or contradicts the hypothesis. In the latter case, that may lead to a modification of the hypothesis, or its complete rejection after which other explanations for observations must then be sought.

If the evidence is supportive, then researchers must decide whether there is sufficient evidence to share their results and interpretation with other scientists.

4. *Undergo peer review.* The sharing of scientific ideas, results, and conclusions takes place through the process of scientific publication in which other independent scientists primarily evaluate whether the evidence supports the conclusions drawn. If and when papers successfully pass the peer review process, they are published in scientific journals.

Reproducibility

A key expectation of the scientific method is that observations, data, and experimental results can be replicated by other competent scientists. Independent replication increases confidence in the evidence supporting a given conclusion. However, if certain evidence cannot be replicated, then doubt may be cast over prior conclusions.

NOTES

Preface

p. xix, Louis got an attack: M. Leakey (1984), pp. 120–121; **"I've got him!":** Morell (1995), p. 181.

p. xx, Interviewer: What are you most: Duncan, D. E. (2003), "Discover Dialogue: Geneticist James Watson." *Discover Magazine*, July 1, 2003. http://discovermagazine.com /2003/jul/featdialogue/.

p. xxi, "As I hope this book": Watson (1969), p. ix.

p. xxi, "The spirit of adventure": Watson (1969), p. ix.

p. xxi, "If history were taught": Kipling (1970).

p. xxi, This has been true: Egan (1989).

p. xxi, It has been claimed: Egan (1989).

p. xxii, People learn from stories: Herman et al. (2010), pp. 473–474.

p. xxii, They have found: Dahlstrom (2014), p. 13615.

p. xxii, For example, it is claimed: Aaker, J., "How to Use Stories to Win Over Others," Lean In, https://leanin.org/education/harnessing-the-power-of-stories.

p. xxii, The process of science: Bruner (1996), pp. 126–127.

p. xxii, "Our instruction in science": Bruner (1996), p. 127.

p. xxii, Even though the advantages: Hadzigeorgiou (2016), p. 91.

p. xxiii, Adam Gopnik: Gopnik, A. (2012). "Can Science Explain Why We Tell Stories?" *The New Yorker*, May 18, 2012. http://www.newyorker.com/books/page-turner/can -science-explain-why-we-tell-stories/.

p. xxiii, "Startle us with": Gopnik, A. (2012). "Can Science Explain Why We Tell Stories?" *The New Yorker*, May 18, 2012. http://www.newyorker.com/books/page -turner/can-science-explain-why-we-tell-stories/.

p. xxv, "I have suffered": Harrison (2007), p. 11.

Chapter 1

p. 2, October 3, 2005: "Nobel Call Interrupts Dinner for Australian Winners," Reuters /abc.net.au, October 3, 2005, http://www.abc.net.au/news/2005-10-03/nobel-call -interrupts-dinner-for-australian-winners/2116712; J. Robin Warren, Interview, "Nobel Prize Talks: J. Robin Warren," Nobelprize.org, May 2, 2014, http://www.nobelprize.org /nobel_prizes/medicine/laureates/2005/warren-interview.html; Barry J. Marshall, Interview, "Nobel Prize Talks: Barry J. Marshall," Nobelprize.org, March 6, 2014, http:// www.nobelprize.org/nobel_prizes/medicine/laureates/2005/marshall-interview.html.

p. 2, "Oh, he is here": J. Robin Warren, Interview, "Nobel Prize Talks: J. Robin Warren," Nobelprize.org, May 2, 2014, http://www.nobelprize.org/nobel_prizes/medicine/lau reates/2005/warren-interview.html.

p. 3, The adventure began: Warren (2002), pp. 151–164.

p. 3, "They appear to be": Warren (2002), p. 155.

p. 3, "If you really believe": Warren (2002), p. 154.

p. 5, Despite the evidence: Marshall (2002), p. 173.

p. 5, After two weeks: Marshall (2002), p. 173.

p. 6, The microorganism should: "Robert Koch, 1843–1910," *Contagion—Historical Views of Diseases and Epidemics: Notable People*, Harvard University Library Open Collections Program, http://ocp.hul.harvard.edu/contagion/koch.html.

p. 7, And just as importantly: Marshall and Warren (1984).

p. 8, The cover letter stated: Marshall (2002), pp. 184–186.

p. 8, "If these bacteria": Marshall and Warren (1983).

p. 10, He experimented with: Marshall (2005), p. 267.

p. 10, Or at the very least: Marshall (2005), p. 269.

p. 11, With a disease-causing: Marshall (2005), p. 268.

p. 11, Marshall chugged: Marshall, Armstrong, McGechie, and Glancy (1985).

p. 11, "Guinea-Pig Doctor": Marshall (2002), pp. 197–198.

p. 12, Marshall wrote up: Marshall et al. (1985).

p. 12, Trials carried out: Unge (2002).

Chapter 2

p. 14, The paper described: Wakefield et al. (1998).

p. 15, "Further investigations": Wakefield et al. (1998).

p. 15, He was worried: Deer (2004).

p. 15, "Measles vaccines are": Offit (2010), p. 22.

p. 15, "One more case": Deer (2004).

p. 15, The British press: Offit (2010), p. 22.

p. 15, "My grandson": Offit (2010), p. 28.

p. 16, "They manufacture products": Offit (2010), p. 30.

p. 16, He revealed that: Offit (2010), p. 31.

p. 16, **Taylor had published:** Taylor, Miller, Farrington, et al. (1999).

p. 16, **"The belief that MMR":** Offit (2010), p. 33.

p. 16, **Wakefield appeared on:** Offit (2010), p. 35.

p. 16, **He was also featured:** Mnookin (2011), pp. 162–163.

p. 16, **"Shame on Officials":** L. Fraser, "Shame on Officials Who Say MMR Is Safe," *Telegraph*, January 21, 2001, http://www.telegraph.co.uk/news/uknews/1318767 /Shame-on-officials-who-say-MMR-is-safe.html.

p. 16, **"We are in the midst":** Brian Deer, "Andrew Wakefield & MMR: How a Worldwide Health Scare Was Launched from London," Brian Deer's MMR Investigation Index, Brian Deer: Selected Investigations & Journalism, http://briandeer.com/mmr/wakefield -archive.htm.

p. 17, **"Our quest for truth":** Brian Deer, "Andrew Wakefield & MMR: How a Worldwide Health Scare Was Launched from London," Brian Deer's MMR Investigation Index, Brian Deer: Selected Investigations & Journalism, http://briandeer.com/mmr/wakefield -archive.htm.

p. 17, **Nor did the study:** Madsen, Hviid, Vestergaard, et al. (2002).

p. 17, **As a precautionary measure:** CDC, "Thimerosal in Vaccines," Vaccine Safety, Centers for Disease Control and Prevention, October 27, 2015, http://www.cdc.gov/vac cinesafety/concerns/thimerosal/.

p. 18, **Magazine articles:** Offit (2010), p. 97.

p. 18, **By 2004, vaccination rates:** G. Thompson, "Measles and MMR Statistics," House of Commons Library Social and General Statistics, Research Briefings, Parliamentary Business, UK, published October 2, 2009, www.parliament.uk.

p. 18, **But again, large-scale:** Hviid, Stellfeld, Wohlfahrt, and Melbye (2003); Immunization Safety Review Committee Board on Health Promotion and Disease Prevention (2004).

p. 19, **His revelations would:** Deer (2004).

p. 19, **Journalist Deer subsequently:** Brian Deer, "MMR Doctor Given Legal Aid Thousands," *Sunday Times*, December 31, 2006, http://briandeer.com/mmr/st-dec-2006.htm.

p. 20, **"Dishonest," "irresponsible," and "misleading":** General Medical Council, "Fitness to Practise Panel Hearing 28 January 2010," pp. 44–47, Brian Deer: Selected Investigations & Journalism, http://briandeer.com/solved/gmc-charge-sheet.pdf.

p. 20, **"Contrary to his duties":** General Medical Council, "Fitness to Practise Panel Hearing 28 January 2010," pp. 54–55.

p. 20, **The board found Wakefield:** General Medical Council, "Determination on Serious Professional Misconduct (SPM) and Sanction," p. 9, Brian Deer: Selected Investigations & Journalism, http://briandeer.com/solved/gmc-wakefield-sentence.pdf.

p. 20, **"An elaborate fraud":** Godlee, Smith, and Marcovitch (2011).

p. 20, **But in 2007–2008:** G. Thompson, "Measles and MMR Statistics," p. 3. House of Commons Library Social and General Statistics, Research Briefings, Parliamentary Business, UK, published October 2, 2009, www.parliament.uk.

p. 20, **Child vaccination rates:** Public Health England, "National MMR Vaccination Catch-Up Programme Announced in Response to Increase in Measles Cases," Press

Release, April 25, 2013, https://www.gov.uk/government/news/national-mmr-vaccination-catch-up-programme-announced-in-response-to-increase-in-measles-cases.

p. 20, **The rate of nonmedical:** Omer, Richards, Ward, and Bednarczyk (2012).

p. 20, **But several individual states:** Constable, Blank, and Caplan (2014).

p. 21, **Since that time:** Fiebelkorn, Redd, Gallagher, et al. (2010).

p. 21, **In 2014, 667 people:** CDC, "Measles Cases and Outbreaks," CDC Measles Home, Centers for Disease Control and Prevention, April 2, 2018, http://www.cdc.gov/measles/cases-outbreaks.html.

p. 21, **That is a frightening:** WHO, "Measles: Fact Sheet," World Health Organization Media Centre, World Health Organization, 2018, http://www.who.int/mediacentre/factsheets/fs286/en/.

p. 21, **A 2015 Gallup poll:** F. Newport, "In U.S., Percentage Saying Vaccines Are Vital Dips Slightly," Gallup, March 6, 2015, http://www.gallup.com/poll/181844/percentage-saying-vaccines-vital-dips-slightly.aspx.

p. 21, **The following table:** Madsen et al. (2002).

p. 22, **In 1964–65, before a vaccine:** CDC, "Measles Cases and Outbreaks," CDC Measles Home, Centers for Disease Control and Prevention, April 2, 2018, http://www.cdc.gov/measles/cases-outbreaks.html.

Chapter 3

p. 24, **For example, possessing:** Allen (1978), pp. 11–12.

p. 24, **"Disorder in the hall":** Allen (1978), p. 13.

pp. 24–25, **"The Star-Spangled Banner":** Allen (1978), p. 8.

p. 25, **Unfortunately for Tom:** Allen (1978), p. 13.

p. 25, **More than 20 years:** Shine and Wrobel (2009), p. 14.

p. 25, **When he was about 10:** Shine and Wrobel (2009), p. 10.

p. 26, **For example, to test:** Shine and Wrobel (2009), p. 36.

p. 26, **"It may be a futile":** Morgan (1903).

p. 26, **Bryn Mawr was:** "A Brief History of Bryn Mawr College," Bryn Mawr College Communications, https://www.brynmawr.edu/about/history.

p. 26, **"Of the graduate":** Brush (1978), p. 171.

p. 27, **In the unfertilized:** Stevens (1905), pp. 11–13.

p. 28, **Stevens speculated:** Stevens (1905), p. 18.

p. 28, **Over the next:** Morgan (1912).

p. 29, **"Two years wasted":** Shine and Wrobel (2009), p. 62.

p. 29, **Virtually all had red eyes:** Of the 1240 offspring, all but three had red eyes. The three white-eyed male flies Morgan obtained in the cross were interpreted to be new mutants (Morgan 1910).

p. 30, **Morgan identified another:** Allen (1978), p. 152.

p. 31, **"These mutations have":** Morgan (1911a), p. 496.

p. 31, **In the fall of 1909:** Sturtevant (1965), p. 46.

p. 31, **Morgan was so impressed:** Shine and Wrobel (2009), p. 82.

p. 31, **Bridges promptly got:** Shine and Wrobel (2009), p. 82.

p. 31, **The workers examined:** Shine and Wrobel (2009), p. 72.

p. 31, **Always frugal:** Sturtevant (1959), p. 298.

p. 32, **Morgan also encouraged:** Allen (1978), pp. 195–196.

p. 32, **"The Boss":** Sturtevant (1959), p. 296.

p. 33, **He went right home:** Sturtevant (1965), p. 47.

p. 33, **"We find coupling":** Morgan (1911b), p. 384.

p. 33, **"The results are":** Morgan (1911b), p. 384.

p. 33, **For example, black body:** Sturtevant (1913).

p. 34, **"The chromosome theory":** Shine and Wrobel (2009), p. 92.

p. 34, **"Miss Stevens had":** Morgan (1912), pp. 468–470.

Chapter 4

p. 36, **More than 60:** R. Thomas, "The Blitz," *West End at War*, Westminster Memories, http://www.westendatwar.org.uk/page_id__152_path__0p2p.aspx.

p. 37, **"For any German":** Letter, Maxted in M. R. Pollock to J. Lederberg, November 9, 1972, The Joshua Lederberg Papers, Profiles in Science, U.S. National Library of Medicine, https://profiles.nlm.nih.gov/ps/access/CCAAKK.pdf.

p. 37, **"We shall not fail":** Letter, Maxted in M. R. Pollock to J. Lederberg, November 9, 1972, The Joshua Lederberg Papers, Profiles in Science, U.S. National Library of Medicine, https://profiles.nlm.nih.gov/ps/access/CCAAKK.pdf.

p. 37, **The weapon had:** "Non-Contact, Parachute Ground (Land) Mine Type GC," Catalogue Number MUN3509, Imperial War Museums, http://www.iwm.org.uk/collections/item/object/30020471.

p. 37, **Just after midnight:** K. Rakerd, "Guildhouse, Belgrave Road SW1," *West End At War*, Westminster Memories, http://www.westendatwar.org.uk/page_id__243.aspx.

p. 37, **He firmly believed:** Lancet (1941).

p. 38, **At the end:** Lancet (1941).

p. 38, *Streptococcus pneumoniae*: Called *Diplococcus* at the time; Brundage and Shanks (2008).

p. 38, **"Ubiquitous and apparently":** Griffith (1922), p. 20.

p. 38, **When bacteria were injected:** Griffith (1923).

p. 38, **This was the first:** Griffith (1928), p. 116.

p. 39, **Rough bacteria lacked:** Griffith (1928), p. 151.

p. 40, **The result was:** Griffith (1928), p. 154.

p. 40, **He found it hard:** Dubos In J. Lederberg, R. Dubos, and M. McCarty, "Symposium Celebrating the Thirty-Fifth Anniversary of the Publication of 'Studies on the Chemical Nature of the Substance Inducing Transformation of Pneumococcal Types,'" The Rockefeller University Caspary Auditorium, February 2, 1979, https://profiles.nlm.nih.gov/ps/access/CCAAOB.pdf.

p. 40, **That was until:** Dubos in J. Lederberg, R. Dubos, and M. McCarty, "Symposium Celebrating the Thirty-Fifth Anniversary of the Publication of 'Studies on the Chemical Nature of the Substance Inducing Transformation of Pneumococcal Types,'" The Rockefeller University Caspary Auditorium, February 2, 1979, https://profiles.nlm.nih.gov/ps/access/CCAAOB.pdf.

p. 42, **The Professor or Fess:** Dubos (1956).

p. 43, **In October 1940:** McCarty in J. Lederberg, R. Dubos, and M. McCarty, "Symposium Celebrating the Thirty-Fifth Anniversary of the Publication of 'Studies on the Chemical Nature of the Substance Inducing Transformation of Pneumococcal Types,'" The Rockefeller University Caspary Auditorium, February 2, 1979, https://profiles.nlm.nih.gov/ps/access/CCAAOB.pdf.

p. 43, **"Disappointment is my":** Fry (2016), p. 71.

p. 44, **For the past:** Letter, O. T. Avery to R. Avery, May 15, 1943, The Oswald T. Avery Collection, Profiles in Science, U.S. National Library of Medicine, https://profiles.nlm.nih.gov/ps/access/CCAABV.pdf.

p. 44, **If we are right:** O. T. Avery to R. Avery, May 15, 1943.

p. 45, **The discovery was:** Avery, MacLeod, and McCarty (1944).

p. 45, **One might think:** Reichard (2002).

p. 45, **The chairman:** H. F. Judson, "No Nobel Prize for Whining," *New York Times*, October 20, 2003.

Chapter 5

p. 48, **Moreover, even several:** Judson (1979), p. 94.

p. 49, **He had avoided the subject:** Judson (1979), p. 47.

p. 49, **"Finding out the secret":** Judson (1979), p. 47.

p. 50, **The features in the picture:** Watson (1980), p. 23.

p. 51, **"Dullest problem imaginable":** Crick (1988), p. 13.

p. 51, **They were impressed:** Watson (1980), p. 34.

p. 52, **Just two years:** Klug (2004), pp. 4–5.

p. 52, **Wilkins told Crick:** Watson (1980), p. 34.

p. 52, **Crick and Wilkins had:** Watson (1980), p. 36.

p. 52, **Franklin was direct:** Maddox (2002), p. 146.

p. 54, **Franklin's X-ray photographs:** Franklin's notes for the seminar indicate her recognition of a spiral (helical) structure (Klug 2004, pp. 12–13), but neither Wilkins nor Watson recalled Franklin addressing the issue of whether DNA had a helical structure (Olby 2009, p. 131).

p. 54, **Later, as Franklin wrestled:** Klug (2004).

p. 54, **He was off:** Ridley (2006), p. 55.

p. 55, **The U.S. government had:** Carroll (2009), pp. 238–239.

p. 56, **Moreover, he had discovered:** Chargaff (1950).

p. 56, **"Their extreme ignorance":** Judson (1979), p. 142.

p. 56, "Well it is all published": Olby (2009), p. 141; Watson (1980), p. 78.

p. 56, "I never met": Judson (1979), p. 142.

p. 56, "A typical British": Judson (1979), p. 142.

p. 56, It immediately occurred: Olby (2009), p. 141.

p. 57, Crick was not convinced: Ridley (2006), p. 62.

p. 57, Franklin, however, continued: Klug (2004), p. 15.

p. 57, Peter received a letter: Watson (1980), p. 91.

p. 57, He snatched the manuscript: Judson (1979), pp. 156–157; Watson (1980), p. 93.

p. 57, "We are still in the game": Watson (1980), p. 94.

p. 58, Franklin turned away: Watson (1980), p. 96.

p. 58, Gosling had obtained: Raymond Gosling, podcast, "Raymond Gosling: DNA Photographer," *NaturePodcast*, April 20, 2013, http://www.nature.com/nature/podcast/index-gosling-2013-04-20.html.

p. 60, When Watson explained: Watson (1980), pp. 111–112.

p. 60, Watson spent the afternoon: Watson (1980), p. 114.

p. 60, Suddenly he realized: Watson (1980), p. 114.

p. 60, "The secret of life": Watson (1980), p. 115.

p. 62, Franklin came to Cambridge: Klug (2004), p. 20.

p. 62, She had concluded: Klug (2004), p. 16.

p. 62, "We all stand": "Due Credit," *Nature* 496, 270 (April 17, 2013), http://www.nature.com/news/due-credit-1.12806; Raymond Gosling, podcast, "Raymond Gosling: DNA Photographer," *NaturePodcast*, April 20, 2013, http://www.nature.com/nature/podcast/index-gosling-2013-04-20.html.

p. 62, "Scientific clowns": Watson (1980), p. 127.

p. 62, "It has not escaped": Watson and Crick (1953).

Chapter 6

p. 64, The affected puppies: Jean Bennett, phone interview with author, June 7, 2017.

p. 64, Lancelot could not: Lewis (2012), p. 247.

p. 65, The solution contained: Acland et al. (2001).

p. 65, And they hoped: Jean Bennett, interview with Rob Whittlesey, July 13, 2016. The team included William Hauswirth from the University of Florida, Gregory Acland and Gus Aguirre from Cornell University, and Albert Maguire and Jean Bennett from Penn.

p. 65, "Jean, these animals": Jean Bennett, interview with Rob Whittlesey, December 17, 2015.

p. 65, "Wow, we can make": Bennett, interview with Rob Whittlesey, December 17, 2015, p. 10.

p. 66, "Simple. Just go": Bennett, interview with Rob Whittlesey, December 17, 2015, p. 1.

p. 67, The feat had been accomplished: See time line in Figure 6.2; "1981–82: First Transgenic Mice and Fruit Flies," NIH National Human Genome Research Institute, last

updated April 26, 2013, https://www.genome.gov/25520307/online-education-kit
-198182-first-transgenic-mice-and-fruit-flies/.

p. 67, "Sure!" Bennett replied: Bennett, interview with Rob Whittlesey, December 17, 2015,
p. 3.

p. 67, After the birth: Bennett, phone interview with author, June 7, 2017.

p. 68, The metabolite is particularly: "Adenosine Deaminase Deficiency," Genetics Home
Reference, NIH U.S. National Library of Medicine, published February 13, 2018,
https://ghr.nlm.nih.gov/condition/adenosine-deaminase-deficiency#genes.

p. 69, Two children: Blaese et al. (1995).

p. 69, "We feel that": Natalie Angier, "Girl, 4, Becomes First Human to Receive Engineered Genes," New York Times, September 15, 1990, http://www.nytimes.com/1990
/09/15/us/girl-4-becomes-first-human-to-receive-engineered-genes.html.

p. 69, "We'd been really": Natalie Angier, "Girl, 4, Becomes First Human to Receive
Engineered Genes," New York Times, September 15, 1990, http://www.nytimes.com
/1990/09/15/us/girl-4-becomes-first-human-to-receive-engineered-genes.html.

p. 70, "I'd hoped initially": Natalie Angier, "Girl, 4, Becomes First Human to Receive
Engineered Genes," New York Times, September 15, 1990, http://www.nytimes.com
/1990/09/15/us/girl-4-becomes-first-human-to-receive-engineered-genes.html.

p. 70, "I have no doubt": Boyce Rensberger, "Cystic Fibrosis Breakthrough," Washington Post, January 10, 1992, https://www.washingtonpost.com/archive/politics/1992
/01/10/cystic-fibrosis-breakthrough/92be2f5a-8e24-4e89-a580-89547d44b79b/?utm
_term=.36055d7024eb.

p. 70, "What we have to show": Boyce Rensberger, "Cystic Fibrosis Breakthrough," Washington Post, January 10, 1992, https://www.washingtonpost.com/archive/politics/1992
/01/10/cystic-fibrosis-breakthrough/92be2f5a-8e24-4e89-a580-89547d44b79b/?utm
_term=.36055d7024eb.

p. 70, The results of Anderson's: Blaese et al. (1995).

p. 70, It was not possible: "Results from First Human Gene Therapy Clinical Trial," NIH
National Human Genome Research Institute, October 19, 1995, last reviewed May 10,
2010, https://www.genome.gov/10000521/1995-release-first-human-gene-therapy
-results/.

p. 70, By the mid-1990s: Villate-Beitia et al. (2015).

p. 70, By 1996, they demonstrated: Bennett et al. (1996).

p. 72, Gelsinger traveled: Steinbrook (2008), pp. 111–120.

p. 72, Just 12 hours later: Lewis (2012), p. 53.

p. 72, Gelsinger died: Lewis (2012), pp. 53–55.

p. 72, Investigators also discovered: Steinbrook (2008), p. 113.

p. 73, "An avoidable tragedy": Sheryl Gay Stolberg, "Senators Press for Answers on
Gene Trials," New York Times, February 2, 2000, http://www.nytimes.com/2000/02
/03/us/senators-press-for-answers-on-gene-trials.html.

p. 73, "Gene Therapy R.I.P.": Tabitha M. Powledge, "Gene Therapy R.I.P.?," Salon,
June 1, 2000, http://www.salon.com/2000/06/01/gene_therapy/.

p. 73, **The cautious:** Bennett (2014).

p. 73, **And very importantly:** Bennett (2014).

p. 73, **Bennett accompanied Lancelot:** Jean Bennett, email to author, June 20, 2017.

p. 74, **"This recent National":** *Congressional Record*, May 23, 2001, v. 147, p. 9224.

p. 74, **Bennett was deluged:** Bennett, phone interview with author, June 7, 2017.

p. 74, **The team treated:** Acland et al. (2005).

p. 74, **"Jean, how would you":** Bennett, interview with Rob Whittlesey, July 13, 2016, p. 8.

p. 74, **High had been working:** Bennett (2014).

p. 74, **Within a few weeks:** Maguire et al. (2008).

p. 75, **Their son would:** Lewis (2012), pp. ix–x, 264–265.

p. 75, **It was like witnessing:** Bennett, interview with Rob Whittlesey, December 17, 2015, pp. 10–11.

p. 76, **For the Phase II:** Bennett, interview with Rob Whittlesey, December 17, 2015, p. 10.

p. 76, **On December 19, 2017:** "FDA Approves Novel Gene Therapy to Treat Patients with a Rare Form of Inherited Vision Loss," U.S. Food and Drug Administration, FDA News Release, December 19, 2017, https://www.fda.gov/NewsEvents/Newsroom /PressAnnouncements/ucm589467.htm.

Chapter 7

p. 78, **On a careful consideration:** Wallace (1908).

p. 78, **A friend who was:** Wallace (1969), pp. 234–235.

p. 78, **By March 12:** Wallace (1969), p. 78.

p. 79, **"When dripping":** Shermer (2002), p. 62.

p. 79, **He was spurred:** Beddall (1969), p. 16.

p. 80, **"The mystery of mysteries":** Toni Vogel Carey, "That Mystery of Mysteries," *Philosophy Now* 105 (November/December 2014), https://philosophynow.org/issues/105 /That_Mystery_of_Mysteries.

p. 80, **"I begin to feel":** Letter, A. R. Wallace to H. W. Bates, 1847, The Natural History Museum, http://wallacefund.info/file/260l.

p. 81, **"I'm afraid the ship's":** Letter, A. R. Wallace to R. Spruce, September 19–October 5, 1852, Natural History Museum General Library, http://www.nhm.ac.uk/research -curation/library/archives/catalogue/dserve.exe?dsqServer=placid&dsqIni=Dserve .ini&dsqApp=Archive&dsqDb=Catalog&dsqSearch=AltRefNo==%27WP1 %2F3%2F24%27&dsqCmd=Show.tcl.

p. 81, **Wallace watched his animals:** Letter, A. R. Wallace to R. Spruce, September 19– October 5, 1852, Natural History Museum General Library, http://www.nhm.ac.uk /research-curation/library/archives/catalogue/dserve.exe?dsqServer=placid&dsqIni =Dserve.ini&dsqApp=Archive&dsqDb=Catalog&dsqSearch=AltRefNo==%27WP1 %2F3%2F24%27&dsqCmd=Show.tcl.

p. 82, **But Professor John:** Van Wyhe (2013).

p. 82, **He brought aboard:** Darwin (1909), p. xv.

p. 82, In September 1832: Keynes (2001), pp. 106–107.

p. 82, "Cargoes of apparent rubbish": Keynes (2001), pp. 106–107.

p. 83, "I know not": Letter, C. Darwin to J. S. Henslow, October 26–November 24, 1832, University of Cambridge Darwin Correspondence Project, http://www.darwinproject.ac.uk/letter/DCP-LETT-192.xml.

p. 83, "If you propose": Letter, J. S. Henslow to C. Darwin, August 31, 1833, University of Cambridge Darwin Correspondence Project, http://www.darwinproject.ac.uk/letter/?docId=letters/DCP-LETT-213.xml.

p. 84, "The stunted trees": Keynes (2001), p. 353.

p. 84, "I loathe, I abhor": Letter, C. Darwin to S. Darwin, August 4, 1836, University of Cambridge Darwin Correspondence Project, http://www.darwinproject.ac.uk/letter/?docId=letters/DCP-LETT-306.xml.

p. 85, "I have specimens": Barlow (1963), p. 262.

p. 85, When I see these: Barlow (1963), p. 262.

p. 86, We may look: All quotes from C. R. Darwin, Notebook B, "Transmutation of Species (1837–1838)," transcribed by Kees Rookmaaker, Darwin Online, http://darwin-online.org.uk/content/frameset?itemID=CUL-DAR121.-&viewtype=text&pageseq=1.

p. 88, "It is clear": Darwin (1839).

p. 88, Two years later: Darwin (1909), p. 52.

p. 88, At last gleams: Letter, C. Darwin to J. D. Hooker, January 11, 1844, University of Cambridge Darwin Correspondence Project, https://www.darwinproject.ac.uk/letter/DCP-LETT-729.xml.

p. 89, On the afternoon: Letter, A. R. Wallace to R. Spruce, September 19–October 5, 1852, Natural History Museum General Library, http://www.nhm.ac.uk/research-curation/library/archives/catalogue/dserve.exe?dsqServer=placid&dsqIni=Dserve.ini&dsqApp=Archive&dsqDb=Catalog&dsqSearch=AltRefNo==%27WP1%2F3%2F24%27&dsqCmd=Show.tcl.

p. 89, "Fifty times since": Letter, A. R. Wallace to R. Spruce, September 19–October 5, 1852, Natural History Museum General Library, http://www.nhm.ac.uk/research-curation/library/archives/catalogue/dserve.exe?dsqServer=placid&dsqIni=Dserve.ini&dsqApp=Archive&dsqDb=Catalog&dsqSearch=AltRefNo==%27WP1%2F3%2F24%27&dsqCmd=Show.tcl.

p. 91, "Received any": Wallace (1855), p. 188.

p. 92, "Why are these": Wallace (1855), p. 190.

p. 92, "To every thoughtful": Wallace (1855), p. 195.

p. 92, "A branching tree": Wallace (1855), p. 191.

p. 92, "It would be difficult": Wallace (1857), p. 481.

p. 93, It has long been reported: George Beccaloni and Charles Smith, "Biography of Wallace," last updated January 2015, The Alfred Russel Wallace Website, http://wallacefund.info/content/biography-wallace.

p. 93, "To think over": Wallace (1905), p. 361.

p. 94, I know not how: Letter, A. R. Wallace to H. W. Bates, December 24, 1860, Natural History Museum, http://www.nhm.ac.uk/research-curation/scientific-resources/collections/library-collections/wallace-letters-online/374/5916/S/details.html;jsessionid=BECF07C5E22E070FD99971396B55B1D0.web-livecluster2#S2.

Chapter 8

p. 96, Only a small portion: Darwin (1869), p. 288.

p. 96, They paused: Yochelson (2001), p. 6.

p. 98, He started out working: Yochelson (1998), p. 10.

p. 99, He figured out: Yochelson (1998), p. 10.

p. 99, By the spring: Yochelson (1998), pp. 10–17.

p. 99, Harvard paid $3,500: Yochelson (1998), p. 33.

p. 99, These first successes: Yochelson (1998), p. 39.

p. 100, His boss reciprocated: Yochelson (1998), p. 85.

p. 100, It was a six-day: Yochelson (1998), pp. 87–94.

p. 101, After more than: Yochelson (1998), pp. 145–149.

p. 101, But unlike his first: Walcott (1883).

p. 101, "Anywhere Walcott cannot": Walcott (1883), p. 152.

p. 102, "To the question": Darwin (1869), p. 307.

p. 102, In the Grand Canyon: Walcott (1883).

p. 102, The fossils he: Walcott (1883).

p. 103, In spring 1894: Yochelson (1998), p. 293.

p. 104, Walcott's quiet way: Yochelson (1998), p. 421.

p. 104, "Tested and tried": Yochelson (1998), p. 399.

p. 105, The summer field trips: "A Voice from the Cambrian," August 1, 2007, Smithsonian, http://www.150.si.edu/chap7/seven.htm.

p. 105, Walcott had never: Yochelson (1998), pp. 48–50.

p. 105, They quickly found: Yochelson (1998), pp. 65–66.

p. 108, Other Burgess animals: Smith (2013).

Chapter 9

p. 111, Any living cell: Delbrück (1949).

p. 111, She was a freshman: Davidson (1999), pp. 66–67.

p. 111, Lynn got the impression: Davidson (1999), p. 69.

p. 112, She was hoping: Properzio (2004).

p. 112, "You did not go": Davidson (1999), p. 81.

p. 113, Her research advisor: Properzio (2004).

p. 114, Lynn enjoyed feeding: Properzio (2004).

p. 114, "May be regarded": Margulis (2005), p. 146.

p. 114, "To many, no doubt": Margulis (2005), p. 146.

p. 114, Moreover, at the time: Sapp (2012), pp. 56–58.

p. 114, **During her first year:** Plaut and Sagan (1958).

p. 115, **Carl received:** Margulis (1998), p. 26.

p. 116, **She was shocked:** Margulis (1998), p. 26.

p. 116, **They also noted:** Ris and Plaut (1962).

p. 116, **"We suggest":** Ris and Plaut (1962), pp. 389–390.

p. 116, **The next year:** Nass and Nass (1963a,b).

p. 116, **And when he was:** Davidson (1999), p. 140.

p. 116, **Carl tried to talk:** Davidson (1999), p. 140.

p. 117, **That did not work:** Sagan (1965).

p. 117, **At Brandeis:** Sagan et al. (1965).

p. 117, **New fossil discoveries:** Barghoorn and Schopf (1966).

p. 119, **When Lynn became pregnant:** Margulis (1998), p. 29.

p. 119, **Finally, the box came:** Margulis (1998), p. 30.

p. 119, **Lynn eventually found:** Margulis (1998), p. 30.

p. 119, **It is never possible:** Margulis (1975), pp. 21–22.

p. 120, **They discovered close:** Zablen et al. (1975); Bonen and Doolittle (1975).

p. 120, **Two years later:** Bonen et al. (1977).

p. 120, **"For her outstanding":** "The President's National Medal of Science: Recipient Details," National Science Foundation, https://www.nsf.gov/od/nms/recip_details.jsp?recip_id=228.

Chapter 10

p. 124, **Brock determined:** Brock and Freeze (1969); Darland et al. (1970); Brock et al. (1972).

p. 125, **They also fanned:** Brock (1967).

p. 125, **"The ultimate scientific goal":** Stanier et al. (1963).

p. 126, **Dear Francis:** Letter, C. Woese to F. Crick, June 24, 1969, Wellcome Trust Woese correspondence archive, https://wellcomelibrary.org/item/b18187870.

p. 126, **The obvious choice:** Letter, C. Woese to F. Crick, June 24, 1969; Wellcome Trust Woese correspondence archive, https://wellcomelibrary.org/item/b18187870.

p. 126, **"I think the project":** Goldenfeld (2014), p. 250.

p. 128, **Woese annotated:** Sapp and Fox (2013).

p. 128, **"Woese, you have":** Woese (2007), p. 3.

p. 129, **The sequences shared:** Zablen et al. (1975).

p. 129, **Woese had encouraged:** Bonen and Doolittle (1975).

p. 129, **They found that:** Bonen et al. (1977).

p. 129, **This molecular evidence:** Bonen et al. (1977).

p. 130, **Not all were absent:** Woese (2007), pp. 1, 6, 7.

p. 131, **Woese learned:** Woese (2007), p. 10.

p. 131, **They suggested:** Woese and Fox (1977).

p. 131, **"A third form of life":** Sapp (2009), pp. 177–178.

p. 131, His colleague Wolfe: Woese (2007), p. 11; Sapp (2009), p. 178.

p. 132, They also wrestled: Woese et al. (1990).

p. 134, Further molecular studies: Zaremba-Niedzwieska et al. (2017).

p. 134, "Opened a door": Otto Kandler, quoted in Sapp and Fox (2013), p. 541.

Chapter 11

p. 135, Pêch Merle Cave: "Pech Merle," http://en.pechmerle.com/the-prehistory-center/the-pech-merle-cave/.

p. 135, Cecilia gashed her head: Leakey (1984), pp. 27–28.

pp. 135–136, In the discarded rubble: Leakey (1984), p. 25.

p. 136, "The best person": Leakey (1984), p. 29.

p. 137, "At least I ended": Leakey (1984), p. 33.

p. 137, Impressed by what: Leakey (1984), pp. 37–39.

p. 139, The climb up: Leakey (1984), p. 54.

p. 139, Starting the descent: Leakey (1984), pp. 54–55.

p. 140, "She and I were": Leakey (1984), p. 59.

p. 140, "It becomes": Leakey (1979), p. 16.

p. 141, "I still am": Morell (1995), p. 98.

p. 141, "Cast a spell": Leakey (1984), p. 63.

p. 142, "When I saw": Leakey (1974), pp. 159–160.

p. 142, Subsequent radiometric dating: Bye et al. (1987).

p. 143, "Never cared": Leakey (1984), p. 98.

p. 145, "I've got him!": Morell (1995), p. 181.

p. 146, "Quite lovely": Morell (1995), p. 183.

p. 147, Louis dashed off a note: L. Leakey (1959), pp. 491–493.

p. 147, But shortly after Zinj: Leakey, Evernden, and Curtis (1961).

p. 148, "Does any animal": Morell (1995), p. 200.

p. 148, This new hominid: Leakey, Tobias, and Napier (1964).

p. 149, Mary's years: Leakey (1966).

pp. 149–150, Those fossils: Morell (1995), p. 447.

p. 150, The buffaloes: Morell (1995), p. 473.

p. 150, These were quickly: Reader (2011), pp. 408–409.

p. 150, Here, Mary: Leakey and Hay (1979).

p. 150, "Now this is": Reader (1981), p. 15.

Chapter 12

p. 153, Bullet holes pocked: Pääbo (2014), p. 29.

p. 154, Once Pääbo learned: Pääbo (1985).

p. 155, When he looked: Pääbo (2014), p. 30.

p. 155, "Raise the hope": Pääbo (1985), p. 645.

p. 156, The method worked: Pääbo (2014), p. 41; Pääbo et al. (1989).

p. 156, For example, Pääbo: Cooper, Mourer-Chauvire, Chambers, et al. (1992).

p. 157, "No, these have not": Pääbo (2014), p. 64.

p. 161, The curator thought: Pääbo (2014), pp. 72–73.

p. 162, Pääbo was already: Pääbo (2014), p. 1.

p. 162, They had done it: Pääbo (2014), p. 1.

p. 163, "A landmark discovery": Lindahl (1997).

p. 163, "It is not within": Lindahl (1997), p. 2.

p. 164, David Reich: Pääbo (2014), pp. 170–171.

p. 164, When the consortium: Pääbo (2014), p. 175.

p. 164, "We now have": Pääbo (2014), p. 176.

p. 165, These two lines: Green, Krause, Briggs, et al. (2010).

p. 165, Genetic evidence suggests: Sankararaman, Patterson, Li, et al. (2012).

p. 165, "The Neanderthals live on": Pääbo (2014), p. 199.

Chapter 13

p. 168, You are who you eat: This chapter was abridged and adapted from *The Serengeti Rules* (Carroll 2016).

p. 168, The large fish eat the small fish: Elton (1927), p. 50.

p. 168, It was a bold and ambitious: Binney (1926), p. 23.

p. 169, He began a diary: C. S. Elton, *Memoir for Royal Society*, unpublished manuscript, Oxford University Bodleian Library Special Collections and Western Manuscripts, Folder A.36.

p. 169, Elton had relatively little time: Elton (1983), p. 5.

p. 169, Elton was, by his own account: Elton (1983), p. 6.

p. 170, "I must do something": Elton (1983), p. 13.

p. 170, With nothing in his stomach: Elton (1983), p. 13; Longstaff (1950), pp. 237–260.

p. 170, Elton dreamed of one day: Stolzenburg (2009), p. 8.

p. 170, Among the most numerous: Summerhayes and Elton (1923); "Bjørnøya," Norwegian Polar Institute, http://www.npolar.no/en/the-arctic/svalbard/bjornoya/.

p. 171, Summerhayes was as keen: Elton (1983), p. 5.

p. 171, He preserved some aquatic animals: Elton (1983), p. 9.

p. 171, "Stew of fulmar": Elton (1983), pp. 23, 30.

p. 171, "Is rather fun": Elton (1983), p. 30.

p. 172, A whale's spout: Elton (1983), p. 34; Gordon (1922), pp. 19–27.

p. 172, Elton had learned that Longstaff: Elton (1983), p. 41.

p. 173, To learn more about how: Elton (1983), pp. 99–100.

p. 173, After more than two months: C. S. Elton, *Small Adventures*, unpublished manuscript, Oxford University Bodleian Library Special Collections and Western Manuscripts, Folder A.32, p. 39.

p. 174, Elton drew a schematic: Summerhayes and Elton (1923).

p. 174, The ship was thwarted: Binney (1926), pp. 24–29.

p. 175, *Norges Pattedyr* **by a Robert Collett:** Biodiversity Heritage Library online, http://www.biodiversitylibrary.org/item/51536#page/9/mode/1up.

p. 175, **Although Elton did not read Norwegian:** C. S. Elton, "How Are the Mice," *Small Adventures*, unpublished manuscript, Oxford University Bodleian Library Special Collections and Western Manuscripts, Folder A.32, p. 1.

p. 175, **There were 50 or so pages:** C. S. Elton, "How Are the Mice," *Small Adventures*, unpublished manuscript, Oxford University Bodleian Library Special Collections and Western Manuscripts, Folder A.32, p. 1.

p. 176, **Like Archimedes in his bathtub:** C. S. Elton, "How Are the Mice," *Small Adventures*, unpublished manuscript, Oxford University Bodleian Library Special Collections and Western Manuscripts, Folder A.32, p. 2.

p. 176, **He also noted that short-eared owls:** Elton (1924), p. 132.

p. 176, **Elton documented the findings:** Elton (1924).

p. 177, **"Chiefly concerned with":** Elton (1927), p. vii.

p. 177, **"It is clear that animals":** Elton (1927), p. 55.

p. 177, **"Subject to economic laws":** Elton (1927), p. viii.

p. 177, **"At first sight":** Elton (1983), p. 55.

p. 177, **"Food is the burning question":** Elton (1927), p. 56.

p. 177, **In Elton's scheme:** Elton (1927), p. 56.

p. 177, **There was generally a progressive:** Elton (1927), p. 69.

p. 178, **"Vast numbers of small herbivorous":** Elton (1927), p. 69.

p. 178, **"All over the world":** Elton (1927), pp. 69–70.

p. 178, **"When a whole lot of lemmings":** Southwood and Clarke (1999), p. 137.

p. 178, **"The lemmings march chiefly":** Elton (1927), p. 133.

p. 178, **The scene was faked:** "The Lemming—A Misunderstood Rodent," BBC Online, http://www.bbc.co.uk/dna/place-lancashire/plain/A43520645.

Chapter 14

p. 180, **One of the persistent challenges:** Peng (1987).

p. 180, **Four Pests Campaign:** Shapiro (2001), pp. 86–89.

p. 181, **"Class, I want you to":** R. Paine, interview, Monterey, California, January 17, 2016.

p. 181, **"Well, what keeps":** R. Paine, interview, Monterey, California, January 17, 2016.

p. 181, **They would soon:** Hairston et al. (1960).

p. 182, **"Why isn't all":** R. Paine, interview, April 1, 2015.

p. 182, **Green world hypothesis:** R. Paine, interview, Monterey, California, January 17, 2016.

p. 183, **That way, the chemist argued:** R. Paine, interview, Monterey, California, January 17, 2016.

p. 183, **"Discovered the Pacific Ocean":** R. Paine, interview, Monterey, California, January 17, 2016.

p. 184, **"There it was":** R. Paine, interview, Monterey, California, January 17, 2016.

p. 184, **"It was clearly":** Stolzenburg (2009), p. 22.

p. 185, The removal of: Paine (1966).

p. 185, "I knew that I hit": R. Paine, interview, Monterey, California, January 17, 2016, p. 12.

p. 185, The mussels advanced: Paine (1974).

p. 185, Within three months: Paine (1974).

p. 185, Over a period: Paine (1971).

p. 186, Just like the stone: Paine (1969).

p. 186, The untouched, control areas: Paine and Vadas (1969).

p. 186, In 1971, Paine: R. Paine, interview, Monterey, California, January 17, 2016.

p. 187, "Have you ever thought": J. Estes and R. Paine, interview by Josh Rosen, Monterey, California, January 15, 2016, p. 4.

p. 187, Their first hint: J. Estes and R. Paine, interview by Josh Rosen, Monterey, California, January 15, 2016, p. 4.

p. 187, "The most dramatic moment": J. Estes and R. Paine, interview, Monterey, California, January 16, 2016.

p. 187, "And that was sticking": J. Estes and R. Paine, interview, Monterey, California, January 16, 2016.

p. 187, "Any fool would": J. Estes and R. Paine, interview, Monterey, California, January 16, 2016.

p. 187, Where otters were: Estes and Palmisano (1974).

p. 187, The kelp was habitat: Estes and Palmisano (1974).

p. 189, Paine coined a new term: Paine (1980).

p. 189, Paine summed up: Paine (2010), p. 25.

p. 189, Once it was realized: Shapiro (2001).

p. 190, We are only now: Estes et al. (2011).

p. 190, For example, the reintroduction: Carroll (2016).

p. 190, *Hyperkeystone*—to describe: Worm and Paine (2016).

Chapter 15

p. 192, When you have seen: Smithsonian Research Reports (1982), p. 44.

p. 192, "How would you like": Wilson (1994), p. 163.

p. 193, "I was scared": Wilson (1991), pp. 36–37.

p. 193, "The longest I've made": Wilson (1991), pp. 36–37.

p. 193, He carried the sticks: Wilson (1991), p. 37.

p. 193, "An unfragmented life": Wilson (1991), p. 41.

p. 194, Ed, don't stay: Wilson (1994), pp. 28–29.

p. 194, "Those from a master": Wilson (1994), p. 29.

p. 194, One day, while fishing: Wilson (1994), p. 13.

p. 195, "Stalking Ants": Wilson (1994), p. 58.

p. 195, "Youth, desire": Wilson (1994), p. 166.

p. 197, **The discovery of new**: Wilson (2010).

p. 197, **Once these immigrants**: Wilson (1959).

p. 197, **Within an encyclopedic**: Darlington (1957).

p. 198, **"Division of area"**: Darlington (1957), p. 483.

p. 199, **Darlington extended this**: Darlington (1971), p. 159.

p. 201, **The data collected**: Quammen (1997), pp. 424–425; https://web.stanford.edu/group/stanfordbirds/text/essays/Island_Biogeography.html; there are some slight discrepancies in numbers in second-hand accounts, but the data generally fit the prediction.

p. 201, *Miniaturize the system*: Wilson (1994), p. 261.

p. 201, **"No one but a naturalist"**: Wilson (1994), p. 263.

p. 202, **"Mine was anything"**: Wilson (1994), p. 263.

p. 202, **Having worked beautifully**: Simberloff and Wilson (1969).

p. 203, **Simberloff later tested**: Simberloff (1976).

p. 203, **Many of the principles**: MacArthur and Wilson (1967), pp. 3–4.

p. 204, **"With the relation"**: Wilson (2016), p. 186.

p. 205, **"Earth, by the 22nd century"**: Wilson (2013), p. 297.

Chapter 16

p. 206, **He wanted instead**: Keeling (1998), p. 30.

p. 206, **Nearly all declined**: Keeling (1998), p. 31.

p. 208, **He was not sure**: Keeling (1998), p. 33.

p. 208, **The air samples**: Keeling (1958).

p. 209, **Newly married**: Keeling (1957); "Revisiting the 'Keeling Curve,'" NPR radio interview, January 28, 2009.

p. 210, **Revelle was not**: Weart (2008), pp. 2–6.

p. 210, **In the 1930s**: Weart (2008), pp. 2–6.

p. 210, **"Thus human beings"**: Revelle and Suess (1957).

p. 211, **"The increase of atmospheric"**: Revelle and Suess (1957).

p. 211, **They had lunch**: Keeling (1998), p. 37.

p. 212, **"We were witnessing"**: Keeling (1998), p. 41.

p. 213, **"This generation has"**: Lyndon B. Johnson, "Special Message to the Congress on Conservation and Restoration of Natural Beauty," February 8, 1965, http://www.presidency.ucsb.edu/ws/?pid=27285.

p. 214, **"This report will"**: Lyndon B. Johnson, "602—Statement by the President in Response to Science Advisory Committee Report on Pollution of Air, Soil, and Waters," November 6, 1965, http://www.presidency.ucsb.edu/ws/?pid=27355.

p. 214, **"Are the single most"**: T. H. Maugh, "Charles David Keeling, 77; Scientist Linked Humans to Increase in Greenhouse Gas," *Los Angeles Times*, June 24, 2005, http://articles.latimes.com/2005/jun/24/local/me-keeling24.

p. 215, **"For his pioneering"**: "Charles D. Keeling," National Medal of Science: Physical Sciences, National Science & Technology Medal Foundation, https://www.national medals.org/laureates/charles-d-keeling#.

p. 215, **By the end of 2016**: "Carbon Dioxide," Global Climate Change: Vital Signs of the Planet, NASA, https://climate.nasa.gov/vital-signs/carbon-dioxide/.

p. 215, **Based on the study**: "Carbon Dioxide at NOAA's Mauna Loa Observatory Reaches New Milestone: Tops 400 ppm," National Oceanic & Atmospheric Administration, May 10, 2013, https://www.esrl.noaa.gov/news/2013/CO2400.html.

p. 215, **The last time global**: "Graphic: Carbon Dioxide Hits New High," Global Climate Change: Vital Signs of the Planet, NASA, https://climate.nasa.gov/climate_resources/7/.

p. 215, **That rate is more**: "Carbon Dioxide at NOAA's Mauna Loa Observatory Reaches New Milestone: Tops 400 ppm," National Oceanic & Atmospheric Administration, May 10, 2013, https://www.esrl.noaa.gov/news/2013/CO2400.html.

p. 216, **Indeed, 16 of the 17**: "Global Temperature," Global Climate Change: Vital Signs of the Planet, NASA, https://climate.nasa.gov/vital-signs/global-temperature/.

p. 217, **"There is a great danger"**: Winston Churchill, speech, April 13, 1939, https://www.raabcollection.com/winston-churchill-autograph/winston-churchill-signed-original-notes-winston-churchills-speech.

Chapter 17

p. 220, **His parents instead promptly**: "Sir John B. Gurdon—Biographical," Nobelprize.org, https://www.nobelprize.org/nobel_prizes/medicine/laureates/2012/gurdon-bio.html.

p. 220, **It has been a disastrous**: "Sir John B. Gurdon—Biographical," Nobelprize.org, https://www.nobelprize.org/nobel_prizes/medicine/laureates/2012/gurdon-bio.html.

p. 221, **Some later Gurdons**: "Sir John B. Gurdon—Biographical," Nobelprize.org, https://www.nobelprize.org/nobel_prizes/medicine/laureates/2012/gurdon-bio.html.

p. 222, **Spemann had found**: Spemann (1938).

p. 222, **They discovered that**: Briggs and King (1952).

p. 222, **However, transferring**: Briggs and King (1957).

p. 224, **But in contrast**: Gurdon (1962).

p. 225, **While he was away**: Gurdon and Uehlinger (1966).

p. 225, **However, it would be**: Gurdon and Byrne (2003).

p. 226, **Since that time**: Gurdon and Byrne (2003).

p. 226, **In 2006, Shinya Yamanaka**: Takahashi and Yamanaka (2006).

p. 226, **Yamanaka's team also**: Takahashi et al. (2007).

p. 227, **In 2012, Yamanaka**: "Shinya Yamanaka—Facts." Nobelprize.org, https://www.nobelprize.org/nobel_prizes/medicine/laureates/2012/yamanaka-facts.html.

Chapter 18

p. 229, **To create is to recombine:** Jacob (1977).

p. 229, **Even before birth:** Aagaard et al. (2014).

p. 229, **They are members:** Lloyd-Price et al. (2016); Findley et al. (2013).

p. 231, **And there were no techniques:** Tonegawa (1988).

p. 233, **This result suggested:** Tonegawa (1988).

p. 233, **Afterward, they cut each:** Hozumi and Tonegawa (1976).

p. 235, **Brack promptly agreed:** Tonegawa (2004).

p. 235, **Wherever along the DNA:** Brack et al. (1978).

p. 238, **"Explain the complex":** Jacob (1977).

Chapter 19

p. 240, **The great pleasure:** "The Split Brain Experiments," NobelPrize.org, October 30, 2003, https://www.nobelprize.org/educational/medicine/split-brain/background.html.

p. 240, **Given any past date:** Treffert and Christensen (2005).

p. 242, **It is estimated:** Treffert (2009).

p. 242, **He reasoned that:** E. E. Smith, "One Head, Two Brains," *The Atlantic*, July 27, 2015, https://www.theatlantic.com/health/archive/2015/07/split-brain-research-sperry-gazzaniga/399290/.

p. 243, **He accessed the corpus callosum:** Mathews et al. (2008).

p. 243, **Despite their severed:** Akelaitis (1944).

p. 243, **Since people appeared:** Sperry (1964).

p. 243, **The paradox attracted:** Sperry (1964).

p. 243, **The cat behaved:** Sperry (1961, 1964).

p. 244, **"It was as though":** Sperry (1964), p. 43.

p. 245, **Then, Gazzaniga headed:** Gazzaniga (2015), pp. 16–17.

p. 245, **The seizures had become:** Gazzaniga et al. (1962).

p. 245, **"You know, even if":** Gazzaniga (2015), p. 39.

p. 246, **In fact, Jenkins said:** Sperry (1964).

p. 246, **"One would hardly suspect":** Sperry (1964).

p. 246, **"What did you see?":** Gazzaniga (2015), pp. 35–36.

p. 248, **He was even unable:** Gazzaniga (2015), pp. 62–63.

p. 248, **"It looked like":** Gazzaniga (2015), p. 63.

p. 248, **The left hemisphere:** E. H. Chudler, "Cerebral Hemispheres," Neuroscience for Kids, https://faculty.washington.edu/chudler/split.html; Gazzinaga (1998).

p. 248, **"Inner world of the brain":** "The Nobel Prize in Physiology or Medicine 1981—Roger W. Sperry, David H. Hubel, Torsten N. Wiesel," NobelPrize.org, October 9, 1981, https://www.nobelprize.org/nobel_prizes/medicine/laureates/1981/press.html.

p. 249, **To prepare for:** "Kim Peek—The Real Rain Man," Wisconsin Medical Society, April 14, 2014, https://www.wisconsinmedicalsociety.org/professional/savant-syndrome /profiles-and-videos/profiles/kim-peek-the-real-rain-man/.

p. 249, **"Special thanks":** "Kim Peek—The Real Rain Man." Wisconsin Medical Society, April 14, 2014, https://www.wisconsinmedicalsociety.org/professional/savant-syndrome /profiles-and-videos/profiles/kim-peek-the-real-rain-man/.

p. 249, **By the time:** B. Weber, "Kim Peek, Inspiration for 'Rain Man,' Dies at 58," *New York Times*, December 26, 2009, http://www.nytimes.com/2009/12/27/us/27peek.html.

BIBLIOGRAPHY

Aagaard, K., J. Ma, K. Antony, et al. (2014). "The Placenta Harbors a Unique Microbiome." *Science Translational Medicine*, 6(237): 237ra65.

Acland, G. M., G. D. Aguirre, J. Bennett, et al. (2005). "Long-Term Restoration of Rod and Cone Vision by Single Dose rAAV-Mediated Gene Transfer to the Retina in a Canine Model of Childhood Blindness." *Molecular Therapy*, 12(6): 1072–1082.

Acland, G. M., G. D. Aguirre, J. Ray, et al. (2001). "Gene Therapy Restores Vision in a Canine Model of Childhood Blindness." *Nature Genetics*, 28: 92–95.

Akelaitis, A. J. (1944). "A Study of Gnosis, Praxis and Language Following Section of the Corpus Callosum and Anterior Commissure." *Journal of Neurosurgery*, 1(2): 94–102.

Allen, G. E. (1978). *Thomas Hunt Morgan: The Man and His Science*. Princeton, NJ: Princeton University Press.

Avery, O. T., C. M. MacLeod, and M. McCarty. (1944). "Studies on the Chemical Nature of the Substance Inducing Transformation of Pneumococcal Types: Induction of Transformation by a Desoxyribonucleic Acid Fraction Isolated from Pneumococcus Type III." *Journal of Experimental Medicine*, 79(2): 137–158.

Barghoorn, E. S. and J. W. Schopf. (1966). "Microorganisms Three Billion Years Old from the Precambrian of South Africa." *Science*, 152(3723): 758–763.

Barlow, N. (1963). "Darwin's Ornithological Notes." *Bulletin of the British Museum (Natural History) Historical Series*, Volume 2, no. 7. London: Trustees of the British Museum.

Beddall, B. G. (1969). *Wallace and Bates in the Tropics: An Introduction to the Theory of Natural Selection*. London: Macmillan.

Bennett, J. (2014). "My Career Path for Developing Gene Therapy for Blinding Diseases: The Importance of Mentors, Collaborators, and Opportunities." *Pioneer Perspectives*, 25: 663–670.

Bennett, J., T. Tanabe, D. Sun, et al. (1996). "Photoreceptor Cell Rescue in Retinal Degeneration (rd) Mice by In Vivo Gene Therapy." *Nature Medicine*, 2(6): 649–654.

Binney, G. (1926). *With Seaplane and Sledge in the Arctic*. New York: George H. Doran Company.

Blaese, R. M., K. W. Culver, A. D. Miller, et al. (1995). "T Lymphocyte-Directed Gene Therapy for ADA SCID: Initial Trial Results after 4 Years." *Science*, 270(5325): 475–480.

Bonen, L., R. S. Cunningham, M. W. Gray, and W. F. Doolittle. (1977). "Wheat Embryo Mitochondrial 18S Ribosomal RNA: Evidence for Its Prokaryotic Nature." *Nucleic Acids Research*, 4(3): 663–671.

Bonen, L. and W. F. Doolittle. (1975). "On the Prokaryotic Nature of Red Algal Chloroplasts." *Proceedings of the National Academy of Sciences*, 72(6): 2310–2314.

Brack, C., M. Hirama, R. Lenhard-Schuller, and S. Tonegawa. (1978). "A Complete Immunoglobulin Gene Is Created by Somatic Recombination." *Cell*, 15: 1–14.

Briggs, R. and T. J. King. (1952). "Transplantation of Living Nuclei from Blastula Cells into Enucleated Frogs' Eggs." *Proceedings of the National Academy of Sciences*, 38(5): 455–463.

Briggs, R. and T. J. King. (1957). "Changes in the Nuclei of Differentiating Endoderm Cells as Revealed by Nuclear Transplantation." *Journal of Morphology*, 100(2): 269–311.

Brock, T. D. (1967). "Life at High Temperatures: Evolutionary, Ecological, and Biochemical Significance of Organisms Living in Hot Springs Is Discussed." *Science*, 158(3804): 1012–1019.

Brock, T. (1998). "Early Days in Yellowstone Microbiology." *American Society for Microbiology News*, 64(3): 137–140.

Brock, T. D., K. M. Brock, R. T. Belly, and R. L. Weiss. (1972). "*Sulfolobus*: A New Genus of Sulfur-Oxidizing Bacteria Living at Low pH and High Temperature." *Archives of Microbiology*, 84: 54–68.

Brock, T. D. and H. Freeze. (1969). "*Thermus aquaticus* gen. n. and sp. n., a Nonsporulating Extreme Thermophile." *Journal of Bacteriology*, 98(1): 289–297.

Brown, D. M. and A. R. Todd. (1952). "Nucleotides. Part X. Some Observations on the Structure and Chemical Behavior of the Nucleic Acids." *Journal of the Chemical Society (Resumed)*, 52–58.

Brundage, J. F. and G. D. Shanks. (2008). "Deaths from Bacterial Pneumonia during 1918–19 Influenza Pandemic." *Emerging Infectious Diseases*, 14(8): 1193–1199.

Bruner, J. (1996). *The Culture of Education*. Cambridge, MA: Harvard University Press.

Brush, S. G. (1978). "Nettie M. Stevens and the Discovery of Sex Determination by Chromosomes." *Isis*, 69(2): 162–172.

Bye, B. A., F. H. Brown, T. E. Cerling, and I. McDougall. (1987). "Increased Age Estimate for the Lower Palaeolithic Hominid Site at Olorgesailie, Kenya." *Nature*, 329: 237–239.

Carroll, S. B. (2009). *Remarkable Creatures: Epic Adventures in the Search for the Origin of Species*. Boston: Houghton Mifflin Harcourt.

Carroll, S. B. (2016). *The Serengeti Rules: The Quest to Discover How Life Works and Why It Matters*. Princeton, NJ: Princeton University Press.

Chargaff, E. (1950). "Chemical Specificity of Nucleic Acids and Mechanism of Their Enzymatic Degradation." *Experiments*, 6: 201–209.

Constable, C., N. R. Blank, and A. L. Caplan. (2014). "Rising Rates of Vaccine Exemptions: Problems with Current Policy and More Promising Remedies." *Vaccine*, 32: 1793–1797.

Cooper, A., C. Mourer-Chauvire, G. K. Chambers, et al. (1992). "Independent Origins of New Zealand Moas and Kiwis." *Proceedings of the National Academy of Sciences*, 89: 8741–8744.

Crick, F. (1988). *What Mad Pursuit: A Personal View of Scientific Discovery*. New York: Basic Books.

Dahlstrom, M. F. (2014). "Using Narratives and Storytelling to Communicate Science with Nonexpert Audiences." *Proceedings of the National Academy of Sciences*, 111(4): 13614–13620.

Darland, G., T. D. Brock, W. Samsonoff, and S. F. Conti. (1970). "A Thermophilic, Acidophilic Mycoplasma Isolated from a Coal Refuse Pile." *Science*, 170(3965): 1416–1418.

Darlington, P. J. (1957). *Zoogeography: The Geological Distribution of Animals*. New York: Wiley.

Darlington, P. J. (1971). "The Carabid Beetles of New Guinea. Part IV. General Considerations; Analysis and History of Fauna; Taxonomic Supplement." *Bulletin of the Museum of Comparative Zoology*, 142(2): 129–197.

Darwin, C. (1837–1838). "Transmutation of Species." Notebook B. CUL-DAR121. Transcribed by Kees Rookmaaker. Darwin Online. http://darwin-online.org.uk/.

Darwin, C. (1838–1841). *The Zoology of the* Beagle, Part 3. Birds. London: Smith, Elder and Co. Darwin Online. http://darwin-online.org.uk/.

Darwin, C. (1839). *Narrative of the Surveying Voyages of His Majesty's Ships Adventure and* Beagle *between the Years 1826 and 1836 Describing Their Examination of the Southern Shores of South America and the* Beagle's *Circumnavigation of the Globe*, Volume III. London: Henry Colburn, Great Marlborough Street.

Darwin, C. R. (1869). *On the Origin of Species by Means of Natural Selection, or The Preservation of Favoured Races in the Struggle for Life*, 5th edition. London: John Murray.

Darwin, C. (1890). *A Naturalist's Voyage Around the World: The Voyage of the H.M.S. Beagle*. New York: Appleton and Co.

Darwin, C. (1909). *The Foundations of the Origin of Species: Two Essays Written in 1842 and 1844*, Francis Darwin (Ed.). Cambridge: Cambridge University Press.

Davidson, K. (1999). *Carl Sagan: A Life*. New York: Wiley.

Day, M. (1976). "Hominid Postcranial Material from Bed I, Olduvai Gorge." In G. I. Issac and E. R. McCown (Eds.), *Human Origins: Louis Leakey and the East African Evidence*. Menlo Park, CA: W. A. Benjamin.

Deer, B. (2004). "Focus: MMR—the Truth Behind the Crisis." *Sunday Times*, February 22, 2004.

Delbrück, M. (1949). "A Physicist Looks at Biology." Quoted in Archibald, J. (2014). *One Plus One Equals One* (p. 88). Oxford: Oxford University Press. Reprinted in Cairns, J., G. S. Stent, and J. D. Watson, eds. (1966). *Phage and the Origins of Molecular Biology*. Cold Spring Harbor, NY: Cold Spring Harbor Laboratory Press. Original source: Delbrück, M. (1949). *The Transactions of the Connecticut Academy of Arts and Sciences*, 38: 173–190.

Dubos, R. J. (1956). "Oswald Theodore Avery. 1877–1955." *Biographical Memoirs of Fellows of the Royal Society*, 2: 35–48.

Egan, K. (1989). "Memory, Imagination, and Learning: Connected by the Story." *Phi Delta Kappan*, 70(6): 455–459.

Elton, C. S. (1924). "Periodic Fluctuations in the Numbers of Animals: Their Causes and Effects." *British Journal of Experimental Biology*, 2: 119–163.

Elton, C. S. (1927). *Animal Ecology*. New York: Macmillan.

Elton, C. S. (1983). "The Oxford University Expedition to Spitsbergen in 1921: An Account, Done in 1978–1983." Norsk Polarinstitutt Bibliotek. Norsk Polarinstitutt, Oslo. http://brage.bibsys.no/xmlui/handle/11250/218913.

Estes, J. A. and J. F. Palmisano. (1974). "Sea Otters: Their Role in Structuring Nearshore Communities." *Science*, 185(4156): 1058–1060.

Estes, J. A., T. Terborgh, J. S. Brashares, et al. (2011). "Trophic Downgrading of Planet Earth." *Science*, 333(6040): 301–306.

Fiebelkorn, A. P., S. B. Redd, K. Gallagher, et al. (2010). "Measles in the United States during the Postelimination Era." *Journal of Infectious Diseases*, 202(10): 1520–1528.

Findley, K., J. Oh, J. Yang, et al. (2013). "Topographic Diversity of Fungal and Bacterial Communities in Human Skin." *Nature*, 498: 367–370.

Fry, M. (2016). *Landmark Experiments in Molecular Biology*. London: Academic Press.

Gazzaniga, M. S. (1998). "The Split Brain Revisited." *Scientific American*, 279(1): 50–55.

Gazzaniga, M. S. (2015). *Tales from Both Sides of the Brain*. New York: HarperCollins.

Gazzaniga, M. S., J. E. Bogen, and R. W. Sperry. (1962). "Some Functional Effects of Sectioning the Cerebral Commissures in Man." *Proceedings of the National Academy of Sciences*, 48(10): 1765–1769.

Godlee, F., J. Smith, and H. Marcovitch. (2011). "Wakefield's Article Linking MMR Vaccine and Autism Was Fraudulent." *British Medical Journal*, 342: 7452.

Goldenfeld, N. (2014). "Looking in the Right Direction." *RNA Biology*, 11(3): 248–253.

Gordon, S. (1922). *Amid Snowy Wastes: Wild Life on the Spitsbergen Archipelago*. New York: Cassell and Company.

Green, R. E., J. Krause, A. W. Briggs, et al. (2010). "A Draft Sequence of the Neandertal Genome." *Science*, 328(5979): 710–722.

Gribaldo, S., A. M. Poole, V. Daubin, et al. (2010). "The Origin of Eukaryotes and Their Relationship with the Archaea: Are We at a Phylogenetic Impasse?" *Nature Reviews Microbiology*, 8: 743–752.

Griffith, F. (1922). "Types of Pneumococci." *Reports to the Local Government Board on Public Health and Medical Subjects*. No 13, pp. 20–45. Ministry of Health. London: His Majesty's Stationery Office.

Griffith, F. (1923). "The Influence of Immune Serum on the Biological Properties of Pneumococci." *Reports of the Local Government Board on Public Health and Medical Subjects*. No. 18, pp. 1–13. Ministry of Health. London: His Majesty's Stationery Office.

Griffith, F. (1928). "The Significance of Pneumococcal Types." *Journal of Hygiene (London)*, 27(2): 113–159.

Gurdon, J. B. (1962). "The Developmental Capacity of Nuclei Taken from Intestinal Epithelium Cells of Feeding Tadpoles." *Development*, 10: 622–640.

Gurdon, J. B. (2013). "The Egg and the Nucleus: A Battle for Supremacy." *Development*, 140: 2449–2456.

Gurdon, J. B. and J. A. Byrne. (2003). "The First Half-Century of Nuclear Transplantation." *Proceedings of the National Academy of Sciences*, 100(14): 8048–8052.

Gurdon, J. B. and V. Uehlinger. (1966). "Fertile Intestine Nuclei." *Nature*, 210: 1240–1241.

Hadzigeorgiou, Y. (2016). "Narrative Thinking and Storytelling in Science." In *Imaginative Science Education*. Basel, Switzerland: Springer International Publishing.

Hairston, N. G., F. E. Smith, and L. B. Slobodkin. (1960). "Community Structure, Population Control, and Competition." *American Naturalist*, 94(879): 421–425.

Harrison, G. (2007). *I, Me, Mine*. San Francisco: Chronicle Books.

Herman, D., M. Jahn, and M. L. Ryan (Eds.). (2010). *Routledge Encyclopedia of Narrative Theory*. New York: Routledge.

Hozumi, N. and S. Tonegawa. (1976). "Evidence for Somatic Rearrangement of Immunoglobulin Genes Coding for Variable and Constant Regions." *Proceedings of the National Academy of Sciences*, 73(10): 3628–3632.

Hugenholtz, P., C. Pitulle, K. L. Hershberger, and N. R. Pace. (1998). "Novel Division Level Bacterial Diversity in a Yellowstone Hot Spring." *Journal of Bacteriology*, 180: 366–376.

Hviid, A, M. Stellfeld, J. Wohlfahrt, and M. Melbye. (2003). "Association between Thimerosal-Containing Vaccine and Autism." *Journal of the American Medical Association*, 290(13): 1763–1766.

Immunization Safety Review Committee Board on Health Promotion and Disease Prevention. (2004) *Immunization Safety Review: Vaccines and Autism*. Institute of Medicine of the National Academies. Washington, DC: National Academies Press.

Intergovernmental Panel on Climate Change. (2001). *Climate Change 2001: The Scientific Basis*. Contribution of Working Group I to the Third Assessment Report of the Intergovernmental Panel on Climate Change, J. T. Houghton, Y. Ding, D. J. Griggs, M. Noguer, P. J. van der Linden, X. Dai, K. Maskell, and C. A. Johnson (Eds.). Cambridge and New York: Cambridge University Press.

Jacob, F. (1977). "Evolution and Tinkering." *Science*, 196(4295): 1161–1166.

Judson, H. F. (1979). *The Eighth Day of Creation: Makers of the Revolution in Biology*. New York: Simon and Schuster.

Keeling, C. D. (1957). "Variations in Concentration and Isotopic Abundances of Atmospheric Carbon Dioxide." In H. Craig (Ed.), *Recent Research in Climatology*. Proceedings of a conference held at Scripps Institution of Oceanography, La Jolla, California, March 25–26, 1957 (pp. 43–49). La Jolla, CA: Committee on Research in Water Resources and University of California.

Keeling, C. D. (1958). "The Concentration and Isotopic Abundances of Atmospheric Carbon Dioxide in Rural Areas." *Geochimica et Cosmochimica Acta*, 13: 322–334.

Keeling, C. D. (1960). "The Concentration and Isotopic Abundances of Carbon Dioxide in the Atmosphere." *Tellus* 12(2): 200–203.

Keeling, C. D. (1998). "Rewards and Penalties of Monitoring the Earth." *Annual Review of Energy and Environment*, 23: 25–82.

Keynes, R. D. (2001). *Charles Darwin's* Beagle *Diary*. Cambridge: Cambridge University Press.

Kipling, R. (1970). *The Collected Works of Rudyard Kipling*. New York: AMS Press.

Klug, A. (2004). "The Discovery of the DNA Double Helix." *Journal of Molecular Biology*, 335: 3–26.

Krause, J., L. Orlando, D. Serre, et al. (2007). "Neanderthals in Central Asia and Siberia." *Nature*, 449: 902–904.

Krings, M., A. Stone, R. W. Schmitz, et al. (1997). "Neandertal DNA Sequences and the Origin of Modern Humans." *Cell*, 90: 19–30.

Lancet. (1941). "Obituary: Frederick Griffith." *Lancet* (May 3, 1941), 237: 588–589.

Leakey, L. S. B. (1959). "A New Fossil Skull from Olduvai." *Nature*, 184: 491–493.

Leakey, L. S. B. (1974). *By the Evidence: Memoirs, 1932–1951*. New York: Harcourt Brace Jovanovich.

Leakey, L. S. B., J. F. Evernden, and G. H. Curtis. (1961). "Age of Bed I, Olduvai Gorge, Tanganyika." *Nature*, 191: 478–479.

Leakey, L. S. B., P. V. Tobias, and J. R. Napier. (1964). "A New Species of the Genus *Homo* from Olduvai Gorge." *Nature*, 202: 7–9.

Leakey, M. D. (1966). "A Review of the Oldowan Culture from Olduvai Gorge, Tanzania." *Nature*, 210: 462–466.

Leakey, M. D. (1979). *Olduvai Gorge: My Search for Early Man*. London: Collins.

Leakey, M. D. (1984). *Disclosing the Past*. Garden City, NY: Doubleday.

Leakey, M. D. and R. L. Hay. (1979). "Pliocene Footprints in the Laetolil Beds at Laetoli, Northern Tanzania." *Nature*, 278: 317–323.

Leakey, R. (1983). *One Life: An Autobiography*. London: Michael Joseph.

Lewis, R. (2012). *The Forever Fix: Gene Therapy and the Boy Who Saved It*. New York: St. Martin's Press.

Lindahl, T. (1997). "Facts and Artifacts of Ancient DNA." *Cell*, 90(1): 1–3.

Lloyd-Price, J., G. Abu-Ali, and C. Huttenhower. (2016). "The Healthy Human Microbiome." *Genome Medicine*, 8(51): 1–11.

Longstaff, T. (1950). *This My Voyage*. New York: Charles Scribner's Sons.

Lydekker, R. (1904). *Library of Natural History*, Volumes II–IV. New York, Akron, and Chicago: Saalfield Publishing Company.

MacArthur, R. H. and E. O. Wilson. (1967). *The Theory of Island Biogeography*. Princeton, NJ, and Oxford: Princeton University Press.

Maddox, B. (2002). *Rosalind Franklin: The Dark Lady of DNA*. New York: HarperCollins.

Madsen, K. M., A. Hviid, M. Vestergaard, et al. (2002). "A Population-Based Study of Measles, Mumps, and Rubella Vaccination and Autism." *New England Journal of Medicine*, 347(19): 1477–1482.

Maguire, A. M., F. Simonelli, E. A. Pierce, et al. (2008). "Safety and Efficacy of Gene Transfer for Leber's Congenital Amaurosis." *New England Journal of Medicine*, 358: 2240–2248.

Margulis, L. (1975). "Symbiotic Theory of the Origin of Eukaryotic Organelles; Criteria for Proof." *Symposia of the Society for Experimental Biology*, 29: 21–38.

Margulis, L. (1998). *Symbiotic Planet: A New Look at Evolution*. New York: Basic Books.

Margulis, L. (2005). "Hans Ris (1914–2004). Genophore, Chromosomes and the Bacterial Origin of Chloroplasts." *International Microbiology*, 8: 145–148.

Marshall, B. J. (2002). "The Discovery That *Helicobacter pylori*, a Spiral Bacterium, Caused Peptic Ulcer Disease." In B. Marshall (Ed.), *Helicobacter Pioneers* (pp. 165–202). Victoria, Australia: Blackwell Science Asia.

Marshall, B. J. (2005). "Helicobacter Connections: Nobel Lecture, December 8, 2005." In K. Grandin (Ed.), *Les Prix Nobel. The Nobel Prizes 2005* (pp. 250–277). Stockholm: Nobel Foundation (2006).

Marshall, B. J., J. A. Armstrong, D. B. McGechie, and R. J. Glancy. (1985). "Attempt to Fulfil Koch's Postulates for Pyloric Campylobacter." *Medical Journal of Australia*, 142: 436–439.

Marshall, B. J. and J. R. Warren. (1983). "Unidentified Curved Bacilli on Gastric Epithelium in Active Chronic Gastritis." *Lancet*, 321(8336): 1273–1275.

Marshall, B. J. and J. R. Warren. (1984). "Unidentified Curved Bacilli in the Stomach of Patients with Gastritis and Peptic Ulceration." *Lancet* 323(8390): 1311–1315.

Mathews, M. S., M. Linskey, and D. K. Binder. (2008). "William P. Van Wagenen and the First Corpus Callostomies for Epilepsy." *Journal of Neurosurgery*, 108: 608–613.

Mellars, P. (2006). "A New Radiocarbon Revolution and the Dispersal of Modern Humans." *Nature*, 439: 931–935.

Mnookin, S. (2011). *The Panic Virus: A True Story of Medicine, Science, and Fear*. New York: Simon and Schuster.

Morell, V. (1995). *Ancestral Passions: The Leakey Family and the Quest for Humankind's Beginnings*. New York: Simon and Schuster.

Morgan, T. H. (1903). "Recent Theories in Regard of the Determination of Sex." *Popular Science Monthly*, 64: 97–116.

Morgan, T. H. (1910). "Sex Limited Inheritance in Drosophila." *Science*, 32(812): 120–122.

Morgan, T. H. (1911a). "The Origin of Nine Wing Mutations in Drosophila." *Science*, 33(848): 496–499.

Morgan, T. H. (1911b). "Random Segregation versus Coupling in Mendelian Inheritance." *Science*, 34(873): 384.

Morgan, T. H. (1912). "The Scientific Work of Miss N. M. Stevens." *Science*, 36(928): 468–470.

Morgan, T. H. (1915). "Localization of the Hereditary Material in the Germ Cells." *Proceedings of the National Academy of Sciences*, 1(7): 420–429.

Nass, M. M. K. and S. Nass. (1963a). "Intramitochondrial Fibers with DNA Characteristics. I. Fixation and Electron Staining Reactions." *Journal of Cell Biology*, 19: 593–611.

Nass, S. and M. M. K. Nass. (1963b). "Intramitochondrial Fibers with DNA Characteristics. II. Enzymatic and Other Hydrolytic Treatments." *Journal of Cell Biology*, 19: 613–629.

Offit, P. A. (2010). *Autism's False Prophets: Bad Science, Risky Medicine, and the Search for a Cure*. New York: Columbia University Press.

Olby, R. (2009). *Francis Crick: Hunter of Life's Secrets*. Cold Spring Harbor, NY: Cold Spring Harbor Laboratory Press.

Omer, S. B., J. L. Richards, M. Ward, and R. A. Bednarczyk. (2012). "Vaccination Policies and Rates of Exemption from Immunization, 2005–2011." *New England Journal of Medicine*, 367(12): 1170–1171.

Pääbo, S. (1985). "Molecular Cloning of Ancient Egyptian Mummy DNA." *Nature*, 314: 644–645.

Pääbo, S. (2014). *Neanderthal Man: In Search of Lost Genomes*. New York: Basic Books.

Pääbo, S., R. G. Higuchi, and A. C. Wilson. (1989). "Ancient DNA and the Polymerase Chain Reaction." *The Journal of Biological Chemistry*, 264(17): 9709–9712.

Paine, R. T. (1966). "Food Web Complexity and Species Diversity." *American Naturalist*, 100(910): 65–75.

Paine, R. T. (1969). "A Note on Trophic Complexity and Community Stability." *American Naturalist*, 103 (929): 91–93.

Paine, R.T. (1971). "A Short-Term Experimental Investigation of Resource Partitioning in a New Zealand Rocky Intertidal Habitat." *Ecology*, 52(6): 1096–1106.

Paine, R. T. (1974). "Intertidal Community Structure: Experimental Studies on the Relationship between a Dominant Competitor and Its Principal Predator." *Oecologia*, 15: 93–120.

Paine, R. T. (1980). "Food Webs: Linkage, Interaction Strength and Community Infrastructure." *Journal of Animal Ecology*, 49(3): 666–685.

Paine, R. T. (2010). "Food Chain Dynamics and Trophic Cascades in Intertidal Habitats." In J. Terborgh and J. A. Estes (Eds.), *Trophic Cascades: Predators, Prey, and the Changing Dynamics of Nature* (pp. 21–36). Washington, DC: Island Press.

Paine, R. T. and R. L. Vadas. (1969). "The Effects of Grazing by Sea Urchins, *Strongylocentrotus* spp., on Benthic Algal Populations." *Limnology and Oceanography*, 14(5): 710–719.

Peng, X. (1987). "Consequences of the Great Leap Forward in China's Provinces." *Population and Development Review*, 13(4): 639–670.

Plaut, W. and L. A. Sagan. (1958). "Incorporation of Thymidine in the Cytoplasm of *Amoeba proteus*." *Journal of Biophysics and Biochemical Cytology*, 4(6): 843–845.

Powledge, T. M. (2000). "Gene Therapy R.I.P.?" *Salon*, June 1, 2000.

Properzio, J. (2004). "Full Speed Ahead." *University of Chicago Magazine*. https://magazine.uchicago.edu/0402/features/speed-print.shtml.

Quammen, D. (1997). *The Song of the Dodo: Island Biogeography in an Age of Extinctions*. New York: Simon & Schuster.

Reader, J. (1981). *Missing Links: The Hunt for Earliest Man*. Boston: Little, Brown, and Company.

Reader, J. (2011). *Missing Links: In Search of Human Origins*. Oxford: Oxford University Press.

Reichard, P. (2002). "Osvald T. Avery and the Nobel Prize in Medicine." *Journal of Biological Chemistry*, 277: 13355–13362.

Revelle, R. and H. E. Suess. (1957). "Carbon Dioxide Exchange between Atmosphere and Ocean and the Question of an Increase of Atmospheric CO_2 during the Past Decades." *Tellus*, 9(1): 18–27.

Ridley, M. (2006). *Francis Crick: Discoverer of the Genetic Code*. New York: HarperCollins.

Ris, H. and W. Plaut. (1962). "Ultrastructure of DNA-Containing Areas in the Chloroplast of *Chlamydomonas*." *Journal of Cell Biology*, 13: 383–391.

Sagan, L. (1965). "An Unusual Pattern of Tritiated Thymidine Incorporation in Euglena." *Journal of Protozoology*, 12(1): 105–109.

Sagan, L. (1967). "On the Origin of Mitosing Cells." *Journal of Theoretical Biology*, 14: 225–274.

Sagan, L., Y. Ben-Shaul, H. T. Epstein, and J. A. Schiff. (1965). "Studies of Chloroplast Development in Euglena XI. Radioautographic Localization of Chloroplast DNA." *Plant Physiology*, 40(6): 1257–1260.

Sankararaman, S., N. Patterson, H. Li, et al. (2012). "The Date of Interbreeding between Neandertals and Modern Humans." *PLOS Genetics*, 8(10): e1002947.

Sapp, J. (2009). *The New Foundations of Evolution: On the Tree of Life*. Oxford: Oxford University Press.

Sapp, J. (2012). "Too Fantastic for Polite Society: A Brief History of Symbiosis Theory." In D. Sagan (Ed.), *Lynn Margulis: The Life and Legacy of a Scientific Rebel* (pp. 54–67). White River Junction, VT: Chelsea Green Publishing.

Sapp, J. and G. E. Fox. (2013). "The Singular Quest for a Universal Tree of Life." *Microbiology and Molecular Biology Reviews*, 77(4): 541–550.

Shapiro, J. (2001). *Mao's War against Nature*. Cambridge: Cambridge University Press.

Shermer, M. (2002). *In Darwin's Shadow: The Life and Science of Alfred Russel Wallace*. Oxford: Oxford University Press.

Shine, I. and S. Wrobel. (2009). *Thomas Hunt Morgan: Pioneer of Genetics*. Lexington, KY: University Press of Kentucky.

Simberloff, D. S. (1976). "Experimental Zoogeography of Islands: Effects of Island Size." *Ecology*, 57(4): 629–648.

Simberloff, D. S. and E. O. Wilson (1969). "Experimental Zoogeography of Islands: The Colonization of Empty Islands." *Ecology*, 50(2): 278–296.

Smith, M. R. (2013). "Ontogeny, Morphology and Taxonomy of the Soft Bodied Cambrian 'Mollusc' *Wiwaxia*." *Palaeontology*, 57(1): 215–229.

Southwood, R. and J. R. Clarke. (1999). "Charles Sutherland Elton 29 March 1900–1 May 1991." *Biographical Memoirs of Fellows of the Royal Society*, 45: 130–146.

Spemann, H. (1938). *Embryonic Development and Induction*. New Haven, CT: Yale University Press.

Sperry, R.W. (1961). "Cerebral Organization and Behavior: The Split Brain Behaves in Many Respects like Two Separate Brains, Providing New Research Possibilities." *Science*, 133(3466): 1749–1757.

Sperry, R.W. (1964). "The Great Cerebral Commissure." *Scientific American*, 210: 42–52.

Stanier, R. Y., M. Doudoroff, and E. A. Adelberg. (1963). *The Microbial World*, 2nd edition. Englewood Cliffs, NJ: Prentice-Hall.

Steinbrook, R. (2008). "The Gelsinger Case." In E. J. Emanuel et al. (Eds.), *The Oxford Textbook of Clinical Research Ethics* (Chapter 10, pp. 110–120). Oxford: Oxford University Press.

Stevens, N. M. (1905). *Studies in Spermatogenesis: With Special Reference to the "Accessory Chromosome."* Washington, DC: Carnegie Institution of Washington.

Stolzenburg, W. (2009). *Where the Wild Things Were: Life, Death, and Ecological Wreckage in a Land of Vanishing Predators*. New York: Bloomsbury.

Sturtevant, A. H. (1913). "A Third Group of Linked Genes in *Drosophila ampelophila*." *Science*, 27(965): 990–992.

Sturtevant, A. H. (1959). *Thomas Hunt Morgan: 1866–1945. A Biographical Memoir* (pp. 283–325). Washington, DC: National Academy of Sciences.

Sturtevant, A. H. (1965). *A History of Genetics*. New York: Harper & Row.

Summerhayes, V. S. and C. S. Elton. (1923). "Contributions to the Ecology of Spitsbergen and Bear Island." *Journal of Ecology*, 11(2): 214–286.

Takahashi, K., K. Tanabe, M. Ohnuki, et al. (2007). "Induction of Pluripotent Stem Cells from Adult Human Fibroblasts by Defined Factors." *Cell*, 131: 861–872.

Takahashi, K. and S. Yamanaka. (2006). "Induction of Pluripotent Stem Cells from Mouse Embryonic and Adult Fibroblast Cultures by Defined Factors." *Cell*, 126: 663–676.

Taylor, B., E. Miller, C. P. Farrington, et al. (1999). "Autism and Measles, Mumps, and Rubella Vaccine: No Epidemiological Evidence for a Causal Association." *Lancet*, 353: 2026–2029.

Tonegawa, S. (1988). "Somatic Generation of Immune Diversity." *In Vitro Cellular & Developmental Biology*, 24(4): 253–265.

Tonegawa, S. (2004). "That Great Time in Basel." *Cell*, S116: S99–S101.

Treffert, D. A. (2009). "The Savant Syndrome: An Extraordinary Condition. A Synopsis: Past, Present, Future." *Philosophical Transactions of the Royal Society B*, 364: 1351–1357.

Treffert, D. A. and D. D. Christensen. (2005). "Inside the Mind of a Savant." *Scientific American*, 293(6): 108–113.

Unge, P. (2002). "*Helicobacter pylori* Treatment in the Past and in the 21st Century." In B. Marshall (Ed.), *Helicobacter Pioneers* (pp. 203–215). Victoria, Australia: Blackwell Science Asia.

Van Wyhe, J. (2013). "'My Appointment Received the Sanction of the Admirality': Why Charles Darwin Really Was the Naturalist on HMS *Beagle*." *Studies in History and Philosophy of Biological and Biomedical Sciences*, 44: 316–326.

Villate-Beitia, I., G. Puras, J. Zarate, et al. (2015). "First Insights into Non-invasive Administration Routes for Non-viral Gene Therapy." In D. Hashad (Ed.), *Gene Therapy—Principles and Challenges* (Chapter 6). InTech. https://www.intechopen.com/books/gene-therapy-principles-and-challenges.

Wakefield, A. J., S. H. Murch, A. Anthony, et al. (1998). "Ileal-lymphoid-nodular Hyperplasia, Non-specific Colitis, and Pervasive Developmental Disorder in Children." *Lancet*, 351: 637–641.

Walcott, C. D. (1883). "Pre-Carboniferous Strata in the Grand Cañon of the Colorado, Arizona." *American Journal of Science*, 26:437–442.

Wallace, A. R. (1855). "On the Law Which Has Regulated the Introduction of New Species." In *The Annals and Magazine of Natural History Including Zoology, Botany, and Geology* (pp. 184–196). London: Taylor and Francis.

Wallace, A. R. (1857). "On the Natural History of the Aru Islands." *Annals and Magazine of Natural History*. Supplement to Vol. XX (December 1857): 473–485.

Wallace, A. R. (1858). "On the Tendency of Varieties to Depart Indefinitely from the Original Type." *Proceedings of the Linnean Society of London*, 3: 53–62.

Wallace, A. R. (1889). *A Narrative of Travels on the Amazon and Rio Negro*. New York: Ward, Lock and Co.

Wallace, A. R. (1905). *My Life*. New York: Dodd, Mead, and Co.

Wallace, A. R. (1908). In *The Darwin–Wallace Celebration Held on Thursday, 1st July, 1908, by the Linnean Society of London*. London: Linnean Society of London.

Wallace, A. R. (1969). *A Narrative of Travels on the Amazon and Rio Negro, with an Account of the Native Tribes, and Observations on the Climate, Geology, and Natural History of the Amazon Valley*. New York: Haskell House.

Warren, J. R. (2002). "The Discovery of *Helicobacter pylori* in Perth, Western Australia." In B. Marshall (Ed.), *Helicobacter Pioneers* (pp. 151–164). Victoria, Australia: Blackwell Science Asia.

Watson, J. D. (1969). *The Double Helix: A Personal Account of the Discovery of the Structure of DNA*. New York: Mentor.

Watson, J. D. (1980). *The Double Helix: A Personal Account of the Discovery of the Structure of DNA* (G. S. Stent, Ed.). New York: W. W. Norton & Company.

Watson, J. D. and F. H. C. Crick. (1953). "Molecular Structure of Nucleic Acids: A Structure for Deoxyribose Nucleic Acid." *Nature*, 171: 737–738.

Weart, S. R. (2008). *The Discovery of Global Warming*. Revised and expanded edition. Cambridge, MA: Harvard University Press.

Wilson, E. O. (1959). "Adaptive Shift and Dispersal in a Tropical Ant Fauna." *Evolution*, 13(1): 122–144.

Wilson, E. O. (1961). "The Nature of the Taxon Cycle in the Melanesian Ant Fauna." *American Naturalist* 95(882): 169–193.

Wilson, E. O. (1991). "Philip Jackson Darlington, Jr. 1904–1983: A Biographical Memoir by Edward O. Wilson." *Biographical Memoirs*. Volume 60. National Academy of Sciences of the United States of America. Washington, DC: National Academy Press

Wilson, E. O. (1994). *Naturalist*. Washington, DC: Island Press.

Wilson, E. O. (2008). Interview. Accessed online: http://www.pbs.org/wgbh/nova /education/activities/3509_eowilson.html.

Wilson, E. O. (2010). "Island Biogeography in the 1960s: Theory and Experiment." In J. B. Losos and R. E. Ricklefs (Eds.), *The Theory of Island Biogeography Revisited* (pp. 1–12). Princeton, NJ: Princeton University Press.

Wilson, E. O. (2013). *The Social Conquest of Earth*. New York, London: Liveright Publishing.

Wilson, E. O. (2016). *Half-Earth: Our Planet's Fight for Life*. New York, London: Liveright Publishing.

Woese, C. R. (2007). "The Birth of the Archaea: A Personal Retrospective." In R. A. Garrett and H.-P. Klenk (Eds.), *Archaea: Evolution, Physiology, and Molecular Biology* (pp. 1–16.). Malden, MA: Blackwell Publishing.

Woese, C. R. and G. E. Fox. (1977). "Phylogenetic Structure of the Prokaryotic Domain: The Primary Kingdoms." *Proceedings of the National Academy of Sciences*, 74(11): 5088–5090.

Woese, C. R., O. Kandler, and M. L. Wheelis. (1990). "Towards a Natural System of Organisms: Proposal for the Domains Archaea, Bacteria, and Eucarya." *Proceedings of the National Academy of Sciences*, 87: 4576–4579.

Wolman (2012). "A Tale of Two Halves." *Nature*, 483: 262.

Worm, B. and R. T. Paine. (2016). "Humans as a Hyperkeystone Species." *Trends in Ecology and Evolution*, 31(8): 600–607.

Yochelson, E. (1998). *Charles Doolittle Walcott, Paleontologist*. Kent, OH: Kent State University Press.

Yochelson, E. (2001). *Smithsonian Institution Secretary, Charles Doolittle Walcott.* Kent, OH: Kent State University Press.

Zablen, L. B., M. S. Kissil, C. R. Woese, and D. E. Buetow. (1975). "Phylogenic Origin of the Chloroplast and Prokaryotic Nature of Its Ribosomal RNA." *Proceedings of the National Academy of Sciences*, 72(6): 2418–2422.

Zaremba-Niedzwiedzka, K., Caceres, E. F., Saw, J. H., et al. (2017). "Asgard Archaea Illuminate the Origin of Eukaryotic Cellular Complexity." *Nature*, 541: 353–358.

ACKNOWLEDGMENTS

This book began over a delicious dinner at a great French restaurant with editors Betsy Twitchell and Jack Repcheck. I was looking forward to working for the first time with Betsy and again with Jack, who shepherded my first books at W. W. Norton, *Endless Forms Most Beautiful* (2005) and *The Making of the Fittest* (2006). But before the project got rolling, Jack, a truly wonderful and beloved man in the prime of life, tragically passed away. We miss him dearly, and we hope that Jack would be pleased with *The Story of Life*.

This book would not have been possible without the contributions and support of many people. Enormous thanks go to Megan Marsh-McGlone, who assisted me throughout the writing and production of the book, assembled the bibliography and notes, and secured scores of figures and permissions. At long last, time to go to Disney World! Special thanks also to Kate Baldwin, who drew many of the figures, and to the late Leanne Olds, who created several illustrations.

I thank the many scientists and family members who shared their stories and gave me access to documents and photos. Special thanks to Robin Warren, Barry Marshall, Jean Bennett, Albert Maguire, Jennifer Margulis, Thomas D. Brock, Richard Leakey, E. O Wilson, and the late Bob Paine and his family. Thanks also to Princeton University Press and Houghton Mifflin Harcourt for allowing me to adapt some stories that were first published in my books *The Serengeti Rules* (2016) and *Remarkable Creatures* (2009), respectively.

I thank my colleagues Drs. Laura Bonetta and Paul Beardsley at the Howard Hughes Medical Institute (HHMI) for their many helpful suggestions, and Dr. Paul Strode for his great work on the Instructor's Guide.

In addition, I received detailed reviews from the following high school teachers and university instructors, and I would like to thank them for their very helpful feedback:

Anne Artz, Preuss School, San Diego, CA
Lisa Ann Blankinship, University of North Alabama
Rebecca Brewer, Troy High School, Troy, MI
Jennifer Broo, Saint Ursula Academy, Cincinnati, OH
Lynn Carpenter, University of Michigan
Dale Casamatta, University of North Florida
Diane Cook, Louisburg College
Gregory A. Dahlem, Northern Kentucky University
Cindy Gay, Steamboat Springs High School, Steamboat Springs, CO
Brenda Leady, University of Toledo
Jennifer Metzler, Ball State University
Bernadine Okoro, District of Columbia Public Schools
Julie Olson, Mitchell High School, Mitchell, SD
Kelly Norton Pipes, Wilkes Early College High School, Wilkesboro, NC
AJ Prawitz, John F. Hodge High School, St. James, MO
Mary Ann Price, Columbia University in the City of New York
Marek Sliwinski, University of Northern Iowa
Anna Bess Sorin, University of Memphis
Melissa Turner, Metea Valley High School, Aurora, IL
Danielle Werts, Golden Valley High School, Santa Clarita, CA

I would also like to thank the team at W. W. Norton for their support and efforts in publishing the book. Special thanks to my editor Betsy Twitchell for her championing of this project, and for all of her input and guidance, and to Connie Parks, Carla Talmadge, Taylere Peterson, Katie Callahan, and Danny Vargo for finding and correcting my mistakes and bringing the book to life. For developing media that will make this book an effective tool in the classroom, I thank media editor Kate Brayton, associate media editor Gina Forsythe, and media editorial assistant Katie Daloia. Megan Schindel, Stacey Stambaugh, Ted Szczepanski, Elyse Rieder, and Patricia Wong are owed a huge debt of gratitude for helping with the permissions for this text. For ably managing the manufacturing of the book, I thank Sean Mintus. I am grateful to marketing manager Stacy Loyal for her advocacy for this book. Thanks also to director of marketing Steve Dunn, director of sales Michael Wright, and all of Norton's talented sales travelers for spreading the word and getting this book into classrooms. Last, I would like to thank Marian Johnson, Julia Reidhead, Roby Harrington, and Drake McFeely for believing in this project.

And finally, very special thanks to my wife, Jamie, for all of the patience, encouragement, and support she has given to my storytelling. This adventure is finished, but our love story continues . . .

INDEX

Note: Material in figures or tables is indicated by italic page numbers.

A

Acland, Greg, 71

Adam's Ancestors (Leakey), 138

adenine, 52, *53*, 56, 60, *61*

adeno-associated virus (AAV), 73, 74

adenosine deaminase (ADA) gene, 68, 69, 70

adenovirus, 70, 71, 72

Aguirre, Gus, 71

Alexander, Lynn. *See* Margulis, Lynn Alexander Sagan

alphaproteobacteria, 134

Amazon River, 79

Amchitka Island, Alaska, 186–87

Anderson, W. French, 66, 68, 69, 70

Animal Ecology (Elton), 168, 176–77, 178

Animal Farm (Orwell), 180, 189

Anomalocaris, 107

antibodies

    antigen-binding site, 230, *232*, 238

    constant (C) regions, 231–32

    disulfide bonds, 230, *232*

    genetic paradox of antibody diversity, 231

    heavy chains, 230, 231, *232*, 238

    joining (J) region, 236

    light chains, 230, 231, 232–33, 236–37, 238

    possible numbers of, 237–38

    production by myeloma cells, 231

    specificity, 229, 230–31

    structure, 230, *232*, 236–37

    and vaccination, 229

    variable (V) regions, 231, *232*

antibody genes

    constant (C) segments, 233, *234*, 235–36, 237

    DNA sequencing, 236–37

    D segments, 237

    genetic paradox of antibody diversity, 231

    germ line theory of antibody diversity, 231, *232*, 233

    heavy chain genes, 237–38

    hybridization with mRNA, 232–33, *234*, 235, *236*

    introns, 236

    joining (J) segments, 236, 237

    light chain genes, 238

    location in myeloma and embryonic cells, 233, *234*, 235–36

    numbers of, 231, 232–33, 237, 238

    rearrangement and recombination, *234*, 235, 237–39

    somatic hypermutation, 238

    somatic theory of antibody diversity, 231, 232, 233

    variable (V) segments, *234*, 235–36, 237–38

antigens, 230–31